One Size Fits None

One Size Fits None

A Farm Girl's Search for the Promise of Regenerative Agriculture

STEPHANIE ANDERSON

University of Nebraska Press | Lincoln and London

Portions of this manuscript originally appeared as "In Search of Lost Grass" in *Kudzu House Quarterly* 6, no. 3/4 (Winter Solstice 2016) and as "The McFarthest Spot" in *Midwestern Gothic* (Summer 2018).

∞

Library of Congress
Cataloging-in-Publication Data
Names: Anderson, Stephanie (Stephanie Renee), 1987– author.
Title: One size fits none: a farm girl's search for the promise of regenerative agriculture / Stephanie Anderson.
Description: Lincoln: University of Nebraska Press, 2019. | Includes bibliographical references.
Identifiers: LCCN 2018011998
ISBN 9781496205056 (pbk.: alk. paper)
ISBN 9781496211927 (epub)
ISBN 9781496211934 (mobi)
ISBN 9781496211941 (pdf)
Subjects: LCSH: Organic farming—United States. | Farmers—United States—Interviews.
Classification: LCC S605.5 .A525 2019 | DDC 631.5/84—dc23 LC record available at https://lccn.loc.gov/2018011998

Set in Garamond Premier Pro
by Mikala R. Kolander.
Designed by L. Auten.

CONTENTS

INTRODUCTION

I'm in western South Dakota, rolling across the prairie in a blue 1970s-era pickup truck, when I first see them. Buffalo—faraway brown dots on a hillside that become massive bodies outside the passenger window as we approach them, their faces accented with beards and curved black horns. They are primeval, ancient, mammothlike. They have a wise look about them, but also a wildness, as when they flash the whites of their eyes, spin around, and gallop off, showing us they'll never be completely tamed.

I'm at Great Plains Buffalo Company, a ranch where Phil and Jill Jerde and their children raise more than a thousand grass-fed buffalo. These buffalo will eventually be slaughtered, providing consumers with meat, but they are much more than food sources. They are the keepers of this grassland. With their hooves they aerate the soil and push seeds into it. With their waste they fertilize it. Through their grazing habits they encourage the growth of grass instead of woody plants. They maintain symbiotic relationships with birds and insects. They make the prairie function in a way it hasn't since their ancestors walked it, before we converted the Great Plains to corn and soybeans.

The buffalo show us what the prairie once was and how humans have changed it—to some, destroyed it—and this in turn is a reminder of all the landscapes we've changed. "Wrong side up," said a Sioux Indian who watched a white sodbuster rip the grassland open with a plow.[1] The Native Americans knew why soil was best left undisturbed: roots, twenty-five miles of them in a single square yard of prairie turf just four inches deep, held the soil in place, had done so for thousands of years.[2] With a single plow swipe the settlers set it free to blow. Result: the Dust Bowl. Later result: desertification turning the Great Plains into a desert.[3] Less than 4 percent of the original tallgrass prairie remains, and those defiant acres are rigorously protected.[4] Still, it is feasible that the tallgrass prairie could be gone before I die. A human being's lifespan is roughly how long it took to destroy 96 percent of it, which does not bode well for the last 4.

But it doesn't have to be this way. The buffalo before me represent a new agriculture that can help restore the prairie and other landscapes without sacrificing the amount of food produced. These animals show us that there are many ways to farm and ranch, that we can change how we define those terms, that we can reverse the damage we have done and create a better agricultural future. The buffalo are walking, breathing proof that human beings do not have to destroy the earth in order to eat.

Years ago, I would not have seen the buffalo as keepers of the range. I grew up about twenty miles from Great Plains Buffalo on a conventional ranch outside of Bison, South Dakota, where my parents raise cattle, wheat, corn, and hay. Had I not discovered a love for writing that drew me to college, I probably would have stayed there the rest of my life, working alongside my father until I could start my own operation. I'm serious about this. Even now, more than ten years after graduating from high school, my "if I had all the money in the world" plan is to buy a ranch somewhere, raise cattle and horses, and write. The ranch I'd run today, though, would be nothing like the ranch my parents run.

We're longtime pals, my father and I. I don't know how many pictures my mother took of me as a kid sitting on his lap in a tractor or in a pickup truck or on an ATV (we call them four-wheelers in South Dakota). Blonde, brown-eyed little me, all smiles, usually gripping the steering wheel pretending to drive, leaning against Dad with his shaggy brown hair, big 1980s glasses, and baseball cap with a cow graphic printed on the front. He taught me to drive a stick-shift pickup at nine, a tractor at twelve, and a swather (a hay-cutting machine) at fourteen. I rode horses on cattle drives and rose before sunrise during calving season to check the pregnant heifers. He taught me almost everything he knows about farming and ranching, lessons I now consider somewhat dubious because, if I wrote them down, they'd form a book on how to farm conventionally, which is also to say industrially.

My dad and I are still pals, don't get me wrong. We just disagree on almost everything about agriculture, though we don't talk much about that. Still, it's a significant rift considering my father's life is the farm. This is not hyperbole. All my father knows is the ranch; he was in his late fifties before he flew on a commercial airplane or waded into the ocean. He seldom meets up with fellow farmers for a beer, and he has not a single hobby. He rarely visits his grown children in their far-flung city apartments. He reads mostly farm-related news, and he did not attend college. I respect his salt-of-the-earth personality, his dedication to his trade, and his strong work ethic, and I know his world is small because he likes it that way. My father doesn't do much besides farming because he simply doesn't want to. That's how much he loves it.

So having his daughter call the type of agriculture he practices into question is a big deal. I'm not trying to embarrass, hurt, or accuse him or anyone who practices conventional agriculture. Quite the opposite. I wrote this book because most conventional farmers and ranchers are good people trapped in a bad system. I believe that beyond a doubt. I respect my father and others like him too much to simply write them off. I'm deeply concerned about their future, because it is my future and yours, too—the world's future for that matter, since the decisions farmers make affect global markets, landscapes, and climates. If we continue farming industrially, then we'll ruin our planet. But if farmers change their practices, we can dramatically increase the odds of reversing climate change. It's time to have a serious discussion about which option they will choose—and that conversation won't be easy, but it's certainly not impossible.

I worked for a farm-and-ranch newspaper in Sioux Falls, South Dakota, right out of college. The biweekly had a total writing staff of two, myself included, so I covered the big stuff right away: the closing of the Sioux Falls Stockyards, the annual Black Hills Stock Show and Rodeo, four state fairs, freezes, floods, and the status of the corn harvest. Some people scoffed at my job—*You write about cows? And corn?*—which only increased my "I'll show 'em" attitude. Before long I was promoted to Special Sections editor. And I got to wear jeans and cowboy boots to work, an undeniable plus.

I was twenty-one and naïve, a good little worker bee, born and bred to believe American agriculture was sacred. My writing, I thought, was a beacon of truth in the lies being spread about farming by the liberal media and the tree-hugging hippies in places like California and New York City. I felt a sense of honor in protecting the farmer and rancher, my heroes, from slander. I sought the facts, which in my mind were as follows: U.S. farmers nobly feed the world by producing nutritious food, protecting the environment, and keeping their rural communities alive. I believed what my sources told me, because my sources were land-grant university professors and state agriculture officials, scientists and county extension service specialists, people my journalism professors had taught me to seek out. Unbiased people, or so I thought. My sources were also farmers and ranchers like my parents, people whose families had farmed the same ground for generations. Good people.

Some stories were harmless, such as profiles of teenage Future Farmers of America (FFA) state officers. Other stories were less innocent. I wrote on the benefits of genetically modified (GM) corn varieties and the latest and greatest machines able to plow, plant, spray, and harvest in record time. I did an especially troubling story about how consumers needn't worry about antibiotic residues in ethanol by-products that are fed to livestock all over the nation. I kept interviewing "family farmers" on "family farms." But I grew increasingly uncomfortable. The tractors, fields, livestock herds, dairies, farm buildings—everything was super-sized, way bigger than what my father and most of our neighbors had. I toured concentrated animal feeding operations (CAFOs), where the cattle looked miserable, and megadairies with sophisticated stainless-steel machines that sucked the milk right out of the udders. I watched massive sprayers douse fields with chemicals. The more I learned about how these farms operated, the more shame and confusion I felt.

I suppose most people have a moment when they realize something they've believed all their life is wrong. For a long time I believed that farmers and ranchers were stewards of the land and that they acted differently than corporate or industrial farms. Now I see that I was part of a powerful agribusiness system glorifying the "progress" of conventional agriculture, a model in which the farm is treated as a factory, industrial farming packaged to look like family farming.[5] My time at the newspaper revealed that everything I thought I knew about farming and food was a lie.

I quit the job at *Tri-State Neighbor* after a year, not only because of my altered perspective but also because I wanted an adventure. I moved to Florida and, a few years later, enrolled in Florida Atlantic University's master of fine arts program to study creative nonfiction. Agriculture was still bothering me, and it showed in my writing: I produced stories set on ranches, elaborate descriptions of cows, a strange tale about a blizzard that kills thousands of livestock. People in Florida asked me about the ranch back in South Dakota and I never knew quite what to say. I didn't know how to talk about agriculture anymore. All I had were questions: *What might a better version of agriculture look like? Are people already doing it, and how might their disparate efforts create a whole, a national conception of what farms should strive to do and how they should think? Is it feasible for conventional farmers and ranchers to switch to something more sustainable? What philosophies should we reject, and what ideas should we hold dear?*

When it came time to select a topic for my MFA thesis, I knew exactly what I had to do: answer the questions. Through extensive research I came to understand the concepts behind regenerative agriculture, but I couldn't visualize them. I had to see them in action, so I traveled to five farms and ranches as part of my quest. I wanted to know what ideas might apply to all farms and how they could be tailored to fit individual environments. On another level I was searching for something very specific: a model for my family's ranch. I needed to know whether we had a place in this better agricultural future.

Because my father believes we probably don't. I can't number the times he's said, "It's too late for us." By this he means it's too late for our ranch to stop being the conventional, industrial operation it is. It's too late for regenerative agriculture. I understand why he believes this. Our farm is big, far too much for him and my brother to handle. He owns a fleet of industrial-sized equipment and has invested all his resources in growing just two cash crops, wheat and cattle. He depends on chemicals and fertilizers and on fattening cattle in a feedlot. He's beholden to operating notes—practically everyone who farms conventionally is. He's afraid that switching to a new system will mean bankruptcy and learning how to farm all over again. This is a fear he shares with people around the world, not just other farmers. What if we try to create a better agriculture and fail?

If I've discovered one thing from the farmers and ranchers in this book, it's this: it is *never* too late to change. Two of them converted existing conventional operations to regenerative ones. Two others entered farming at middle age, starting from scratch with no experience. People like this show us change is always possible, and it's not as hard as we think. The changes described in this book are not revolutionary or new. They are about returning to time-tested philosophies and perspectives about growing food and reimagining them for the modern world.

One argument of *One Size Fits None* is this: it is time for agriculture to go beyond "sustainable," the food and farming buzzword of the last decade. "Sustainable" has long been the battle cry of books and blogs, and in theory the term means agriculture that returns the resources it takes from the earth, whether through biological practices or careful use of nonrenewables. But the farmers and ranchers I interviewed say this "give back what you take" approach does not do enough to restore soil health, recharge grasslands, fend off desertification, or provide nutrient-rich food for consumers. That is what Gabe Brown, a North Dakota farmer, told me as we bounced over a field in his Polaris Ranger on our way to see some soils. "That's the cliché word; everybody wants to be sustainable," he said that day. "But why do you want to sustain a degraded resource? We need to be regenerative. If we are going to have healthy food and healthy soils for the next generation, and generations to follow, we've got to build our soils back." Given that much of our nation's farm and ranch land is already degraded, sustainable agriculture often means maintaining a less-than-ideal status quo. Industrial agriculture has also co-opted the term for marketing purposes without implementing better farming and animal production practices. Even Monsanto, the company that first created genetically modified crops and sells dozens of harmful agrochemicals, claims to be "a sustainable agriculture company" and produces an annual sustainability report.[6] Unfortunately many "sustainable" farms and ranches are just large-scale conventional operations in disguise. Regenerative agriculture, in contrast, creates new life and resources—and it is already leading the next wave of green food production.

Such agriculture only works, however, when farmers tailor its implementation to their local environment. There is no one-size-fits-all approach in practicing it, and that's another argument in this book. We have used a one-size model for the last sixty or so years, which is the "get big or get out," input-focused philosophy known as conventional agriculture. As we continue the transition away from conventional farming, we must be careful not to think of regenerative agriculture as a specific set of practices or rules. We need to avoid the one-size-fits-all thinking that got us into trouble and instead embrace the idea that one size fits none. Even large-scale farms, so often vilified, can play a role in the new regenerative agriculture, as can super-small urban farms. Midsize farms, however, will likely be the backbone of a regenerative agricultural system.

I turn to farmers and ranchers for evidence. *One Size Fits None* starts with Ryan Roth, a conventional vegetable and sugarcane farmer in Florida whose story reveals how producers became mired in the conventional system in the first place, to the detriment of their finances, the environment, and their communities. This section explains why conventional agriculture is no longer feasible, even when combined with sustainable practices. The book continues with Phil Jerde, the South Dakota buffalo rancher using a form of regenerative agriculture called holistic management and an unexpected animal, the buffalo, to restore native prairie and provide an alternative to the industrial feedlot. Phil proves that livestock are the cornerstone of regenerative agriculture in many environments and that, in some cases, large-scale farms actually work. From there, Kevin O'Dare of Florida and Fidel Gonzalez of New Mexico show how regenerative agriculture that is organic, local, and urban can revitalize communities. Kevin and Fidel also highlight the positive role of super-small farmers who defy the conventional logic of "get big or get out" and have the unique ability to feed their neighbors. In the book's last section, midsize farmer Gabe Brown combines livestock and grain farming to revive soil, heal grassland, feed the local community, and bring back wildlife—regenerative agriculture in diversified form. Gabe also trains the next generation of farmers through his internship program, and he spends half the year on a speaking tour spreading the word about regenerative agriculture. These farmers tailor their operations to their specific environments and goals, proving that, when it comes to regenerative agriculture, one size fits none—an approach that offers more life-giving benefits for the consumer, land, livestock, community, and farmer than sustainable agriculture does in its current form.

I capture a roughly one-year span in these farmers' lives through on-farm visits supplemented with follow-up interviews via email and over the phone. I also provide a history of their operations. Since those conversations several years ago, they have made changes, expansions, and improvements. Due to the nature of book publishing, it's impossible to have an up-to-the-minute picture of anyone's life—but what follows is what I saw during my time with them, and their stories are as authentic now as they were then.

All this brings me back to the buffalo.

As I watch the animals graze, I imagine that I'm on the prairie of old, before white settlers arrived, when buffalo marched in herds thousands strong. They remind me that nearly extinct things can be resurrected. We can make these resurrected things new, retool them for a different time, as Phil has with his herd. The buffalo also call us to remember what we've forgotten, and they promise that if we do we can recover some of what we've lost.

They are all this, yet more—they are a symbol for the future. They represent the variety we need in our diets and our agriculture, the end of monocultures and the revival of diversity. They stand for the use of ecosystems, not chemical solutions, to grow food. They remind us of the tools nature provides, tools that are free and regenerative. Buffalo almost disappeared from the earth because we wanted to farm the prairie they lived on. Now the buffalo gazing at me through the pickup window embody an agricultural change that's already started.

PART ONE
Conventional

1

The Vice President

"Is that a fire?"

From the passenger seat of Ryan Roth's pickup truck, I point to a brownish-black plume to the north. The haze drifts lazily upward, then grows into a broad cloud that resembles a tornado barreling over a freshly plowed field, the kind of swirling brown mass that eats up Kansas towns. Ryan explains that sugarcane must be burned before harvesting. Sugarcane is the number one crop here in the Everglades Agricultural Area (EAA), a seven-hundred-thousand-acre stretch of farmland south of Florida's Lake Okeechobee.[1] Growers can burn a forty-acre block of sugarcane (*block* means *field* in this part of the country) in about five to ten minutes. Greener cane might take twenty. Burning removes the trash, or the leaves and tops, so that the stalks are bare and easier to harvest. Harvesting machines, called combines, drive into the charred cane, cut it at the base, strip away any remaining trash, and chop it into pieces that are hauled to the sugar mill for processing.

I comment that it seems brave to burn in such strong winds. Prairie fires, good as they are for the health of the grassland, were always a danger when I was growing up in South Dakota, something that sticks with me still. You never, *ever* lit a fire when it was windy. Ryan chuckles and says, "The only thing that will stop them from burning is if the wind is coming out of the northwest or out of due west because then all the ash will fall on Wellington, and people will complain they have ash on their cars."

Wellington is a town less than thirty miles from here, an international destination for equestrian events like polo, show jumping, and dressage. Wellington's luxury cars and the people who drive them outweigh Belle Glade's harvest schedules and farmers.

Throughout the morning, smoke clouds ignite, fume furiously, and then vanish, some less than a mile away as we drive, others far in the distance, all carrying a smell like charcoal. The fires seem sporadic at first, but Ryan tells me that every fire requires a permit beforehand, so at least the fire department, if not me, is prepared for the blaze. Sometimes three or four burn at once. They make me uneasy. Ryan, however, is unmoved. The fires are a good metaphor for his life, at least during the farm's busy months, from September to May, growing season in southern Florida. He constantly deals with problems that blaze up out of nowhere. As soon as he puts out one fire, another flashes to life.

His cell phone's ring is the fire alarm. It rings roughly every five minutes, sometimes more frequently. The first call comes from the packinghouse; the speaker complains that the radishes Ryan sent this morning have black spot, a disease that gives them dark splotches. About 10 percent of the radishes rolling down the grading line have it, the speaker says. Shoppers won't buy radishes with even a hint of black spot, so they'll have to be thrown away, a loss for the farm. Ryan says 10 percent is not too bad and asks if the packinghouse can continue removing the diseased ones, which the speaker agrees to. Ryan asks the speaker to inspect the next load carefully and let him know what percentage has spots; if it's too high, he won't pull (the local term for "harvest") any more today.

Ryan hangs up and starts to apologize for being on the phone, but another call interrupts him, this time from the mechanic saying a machine has broken down. Not long after, someone else rings in to see if Ryan needs a crew to hand-cut seed cane, and someone else checks in after that about lettuce boxes. Of course the boxes are wrong. They aren't reinforced as Ryan ordered and the lettering on the label isn't right. Sometimes Ryan will be on the phone and get another call, which he'll have to ignore and return later.

Ryan has just a hint of a southern accent. He says "please" and "thank you" and inquires how people are before listening to their bad news. He speaks with precision; I can tell he is trying to make sure no one misunderstands him, because a misunderstanding can cost thousands of dollars. Once he says the word "shit" and immediately covers the phone with his hand and whispers "sorry." I say it's fine. I imagine I'd be cursing much more if I received calls at this rate. Ryan seems stressed and weary. When he's not talking on it, he clips his phone to the waist of his pants or sets it in the pickup's center console until it once again explodes into its cheery ringtone at maximum volume.

Given the metaphorical fires, I'm grateful that Ryan is taking time to drive me around. A lot of farmers wouldn't during harvest season. Today I'm here to see workers pull radishes and lettuce and hand-cut seed cane on Roth Farms, where Ryan is the vice president and grandson of the farm's founder. Roth Farms spans 3,500 acres that were once six individual farms.[2] Roughly 2,700 of those acres, most of the farm, are committed to sugarcane. The rest support vegetables and leaf, a term denoting any leafy crop such as lettuce and cabbage. Ryan grows rice during the off-season, not for profit but as a cover crop to hold soil and water in place and control weeds. Here's the rotation for a typical block: four years of sugarcane, one growing season of vegetables and leaf, one summer of rice, another growing season of vegetables and leaf, then four years of cane. Repeat.[3]

I found Ryan through the county extension agent. I asked the agent if he could introduce me to a small, local farmer to interview for my master's thesis. I drove out to Belle Glade for a meeting of the Lettuce Advisory Committee, a group of local lettuce growers, where I would meet the farmer. The committee was started in response to a nasty virus that invaded the fields years ago. The group's focus has since shifted to downy mildew, a leaf disease. Downy mildew "spreads like wildfire," as the agent put it. But the meeting was essentially a platform for representatives from chemical companies like Bayer and Helena to advertise sprays for downy mildew. Growers get to choose their own spray, but it's implied that everybody—*everybody*—must spray. After the meeting, I met Ryan, who told me a little about his operation and agreed to host me at the farm in a few months. That's when I realized my understanding of "small, local farmer" was much different than the extension agent's. Still, I sensed Ryan had a story and a perspective worth exploring—and I was right.

Ryan pulls the truck over and we walk into a radish field where workers are gathered at a harvester. Rounded red tops peek out of the soil. Our footsteps bury the radishes under the dirt, derailing a month of growth. I feel guilty about pushing them out of the harvester's reach with my cowboy boots, essentially taking dollars from Ryan's pocket. In my head, I hear my father's voice warning my sisters and I not to ride our horses in the wheat and corn fields—he didn't want hooves crushing the stalks—but Ryan says nothing. I learn later that Roth Farms makes very little money on radishes, so maybe that accounts for his indifference.

Every week from the beginning of October to the end of April, Roth Farms plants thirty to seventy new acres of *Raphanus sativus*, a member of the Brassicaceae family. At about half a million radish seeds per acre, that's 650 million radishes planted in seven months. Crews planted the radishes I'm walking on the first week of October. Roughly thirty-two days later, today, they are ready for harvest. Ryan yanks three radishes out of the black soil and studies the pinkish-red bulbs, frowning. A worker tells Ryan he's "seeing a lotta black spot," at which Ryan frowns deeper. "Black spot has been an ongoing issue for the last four to five years," Ryan says with an exasperated sigh. Black spot isn't the only problem: he pulls a green cutworm off his shirt and, holding it for a moment between his finger and thumb, says, "We got worms out here, too." He tosses the wriggling worm into the field.

Ryan finishes discussing the black spot situation and the men fire up the harvester, which is actually an old corn combine reworked twenty-five years ago with a homemade, radish-appropriate harvesting head. Ryan says they baby the machine because it's old and also because it's almost impossible to buy a new radish harvester, because radishes are not a popular crop in the U.S. anymore. The machine smells of grease and diesel fuel, much like the wheat harvesters I grew up watching in South Dakota. Ryan and I walk backward in front of the harvester's whirring head, and he explains how the spinning gears grab each radish by its green top and pluck it out of the ground. The machine feeds the dangling radish up into the head, where a blade slices off the green top. The round, ruddy root then travels out of the head and plops into a waiting wagon, where it will ride several miles to the packinghouse in Belle Glade and join thousands of other radishes on the grading line.

At twenty-nine, Ryan is three years my senior, and he has worked on his family's farm fulltime since 2005. He reminds me of the young farmers I met in eastern South Dakota while working for *Tri-State Neighbor*—energetic, tech-savvy, eager to talk. During the high season, from January to April, Ryan is on the farm seven days a week. From May to August, the rainy season, he scales back to five days. He bumps up to six days a week from September to December. A day is at least eight hours, often ten, sometimes more. As the farm's vice president, Ryan oversees almost everything: machine-shop activities; maintenance of farm infrastructure such as roads and water pumps; land cultivation and water control; and the planting, spraying, general care, and harvesting of all crops—sugarcane, leafy vegetables, radishes, celery, and rice.[4] He is responsible for these tasks not in the sense that he completes them personally, but in that he manages the employees who do and provides them with guidance and instruction.

His job used to involve more physical work, which Ryan has reluctantly stopped doing so he can spend more time coordinating farm tasks. "I come home really dirty from work when I decide to, more than because I have to. That's not a good thing," he says. "I like work; it used to be where it was raining and there'd be lettuce in the ground and a lot of the ditches needed to be shoveled to make sure the water was leaving the field. And

I'd be out there shoveling with them. Now I still do that from time to time, but I do that once all the decisions have been made. It was hard for me to break myself of that."

Overseeing a farm this size is not easy. The farm employs harvesting and planting crews and other as-needed workers—about 130 people over the course of a year. About 25 to 30 separate employees are considered permanent; they stay year-round except for the months of June and July, work only for Roth Farms, and have 401Ks, health insurance, vacation pay, holiday pay, and a salary. Another 4 to 5 employees stay completely year-round with similar benefits. Ryan somehow manages the activities of this large staff while balancing his other farm duties. Judging from the rate at which his phone rings, he needs an assistant.

Though Ryan never drives a tractor or plants a seedling, he is considered a family farmer, and Roth Farms is considered a family farm. The United States Department of Agriculture (USDA) defines a family farm as "any farm where the majority of the business is owned by the operator and individuals related to the operator by blood or marriage, including relatives who do not reside in the operator's household."[5] Nonfamily corporations or cooperatives do not count. Under this definition virtually all U.S. farms are family owned (97 percent).[6] A closer look at the numbers reveals that, when it comes to the farms producing the majority of the food we eat, being "family owned" is not as meaningful as it sounds.

The Economic Research Service (ERS) of the USDA classifies family farms according to gross cash farm income (GCFI). GCFI includes the farm's sales of crops and livestock, government payments, and other farm-related income. The ERS classifies family farms into three sizes: small (GCFI of less than $350,000), midsize ($350,000 to $999,999) and large ($1 million or more). Ninety percent of all U.S. farms are small family farms—a nice-sounding statistic—but this sector only produces 26 percent of the value of U.S. agricultural production.[7]

Large-scale and midsize family farms, on the other hand, make up almost 8 percent of all farms, but they produce 60 percent of U.S. farm output (25 percent for midsize, 35 percent for large-scale farms). The 3 percent of U.S. farms that are nonfamily corporate farms generate 15 percent of the nation's farm output. In other words, these large-scale and midsize family farms and nonfamily farms are so gigantic that together they produce 75 percent of what we eat, even though they make up only 10 percent of the total number of farms—a vulnerable, top-heavy system.[8]

The large-scale and midsize family farms feeding us are overwhelmingly industrial operations where conventional tools such as agrochemicals, synthetic fertilizers, and genetically modified crops are the norm. These farms often cover thousands of acres and operate like corporations. Roth Farms is one of these, complete with a corporate-style structure. Ryan's dad, Rick, and three aunts own Roth Farms, but Rick's sisters "do not make farm decisions," as Ryan puts it. Rick holds the title of president. An uncle, the husband of one of his dad's sisters, is the farm manager—*was* the farm manager, until he was caught embezzling money. The fraud was discovered about six weeks before my visit to see the radishes.

The words "family farm" might conjure idyllic images of pastoral life—children cupping yellow chicks or wandering in green pastures—but the farm was not a big part of Ryan's childhood, a fact that is surprising to someone like me who grew up making friends with cows and riding in tractors. Our house is steps away from the barn, pastures, and wheat fields. Deer graze the yard at night and coyotes howl at dusk (they even did so during my outdoor ranch wedding service, much to everyone's amusement). The ranch shaped me, wove me into its rhythms, and made me part of its landscape. I can't fathom being a farmer's child without knowing the farm. Ryan, however, grew up in Wellington and his father commuted to the farm. "I knew that he worked on a farm and I came out a few times, but I guess I never really put two and two together that my dad owned it," Ryan says when I ask about his younger years. "I just thought my dad worked on a farm."

Ryan visited the farm more as he grew older, although as a teenager he didn't find farm work as compelling as I did at that age. When he was thirteen, Rick set him to mowing six hundred acres of turfgrass, a crop the farm used to grow, during one of Florida's hot, humid summers. Ryan is actually allergic to grass. Ten hours a day, five days a week, sneezing, coughing, eyes watering—this was Ryan's inaugural farmwork experience. It lasted six weeks. The next summer he found a different job. At fifteen, he returned to work in the farm's machine shop rebuilding a radish harvester, which he says was about as tedious as mowing. "I said, 'You know what, I'm never coming back here. I don't care, I just never want to work on a farm again.'"

How in the world did you end up farming? I ask. Ryan tells me that during his sophomore year at Florida State University in Tallahassee he took his "first real business class" as a business major. The professor asked the students to write a paper in which they imagined themselves at sixty-five years old. The professor wanted to know: Where are you in your career? Are you running a company? Are you working for someone else or for yourself? If you could design your career perfectly, what would it look like?

"I had no answer," Ryan says. "Damn it, I didn't have a passion. I don't have one specific industry that I'm passionate about. I feel like I can coordinate people, I feel like I'm good at getting the most and the best out of people. I feel like for whatever reason people have wanted to—I don't know if the right word is follow me or do what I say or what—but I've been able to get things done. So what am I passionate about? I don't know. I just feel like I'd be good at running a company. Shoot, at that point, does it matter where you work? I didn't want to work on the farm, but it was my best shot at actually being able to run a company. I could go to work for anybody and, if I never got recognized or if I screwed up or whatever, I may never have the opportunity to ever run a company. So I transferred to the University of Florida, got an agriculture degree, and came back here. As long as I don't screw up, I hopefully have a chance to run a company."

I ask Ryan how he describes himself. Does he say "I am a farmer"? Or does he say "I manage a farm" or "I run a company"? Something else? To me, the words "farmer" and "rancher" imply a special identity, a stewardship and an almost spiritual sense of responsibility for the land. A farmer or rancher believes there is no other work that will satisfy an inner desire to coax life and food from the soil. Where that desire comes from isn't always clear. It's like the calling I feel to write, or that others feel to be chefs, priests, and parents—it's insistent and powerful. It's also a throwback to the "agrarian myth" of old, created by early American writers like J. Hector St. John de Crèvecœur, who suggested that farming was not simply an occupation, but a way of thinking and living. The yeoman farmer (meaning a small, landowning family farmer) was, according to Thomas Jefferson, the backbone of American democracy and the epitome of republican values. In a letter to John Jay, we see Jefferson's views clearly: "Cultivators of the earth are the most valuable citizens. They are the most vigorous, the most independent, the most virtuous, and they are tied to their country, and wedded to its liberty and interests, by the most lasting bonds."[9]

With this farmer identity comes an intimate understanding of and appreciation for the land, and the desire to preserve it. This identity makes the struggles of farming—weather, finances, crop failures—worth it, because the farmer so thoroughly loves the role of steward. The work provides something more meaningful than money: a way of living that nourishes the human spirit. This feeling also makes the task noble, a sacrifice for the good of humanity, country, and the environment.

This is how I understood the words "farmer" and "rancher" *before* I understood the industrial farming system and the fact that most food producers see their work simply as a job, not a way of being. I no longer believe in the presence of the farmer and rancher identity in American farm country, not because it doesn't exist (in many people, it does) but because I realize it is not required of modern farmers and is therefore felt by fewer of them. Those who do recognize the importance of this identity are encouraged to ignore it because the components of stewardship—land, flora and fauna, water, nutritious food, healthy soil, community—have little or no place in the business world.

The modern farmer's agricultural degree teaches farming as a business, not an identity. Today's farmer is someone who runs a company, a business-minded "people person" rather than a plant, soil, or livestock person. Conventional farmers can, if they want to, harvest an entire crop without once touching the soil. The farmer isn't even a farmer but a "grower" or "producer," terms that reflect modern agriculture's obsession with yield. Those names also put the farmer in his or her place on the food assembly line. The farmer's one job is to produce, to increase yield per acre.

Modern farming has erased stewardship. Agricultural writer Lisa Hamilton describes the problem like this: "As artificial inputs and external supports have become integral parts of our food production, they have eliminated the farmer—the steward. Of course farmers are still called upon to perform the perfunctory acts of buying seeds and applying fertilizer and driving machinery, but no longer are they asked to think or care or protect."[10] Today's farmer is concerned primarily with inputs versus outputs, a formula that doesn't take environmental, social, and health costs into account. Hamilton's observation echoes Wendell Berry's, who writes that when farmers industrialize, they forsake the role of nurturer and take on the role of technician or businessperson.[11] Being a farmer, he says, involves complexity of thinking, culture, and character, whereas being a businessperson or technician often simplifies and destroys those things. If the farmer simplifies agriculture to expenses against profits, then there is no room or reason for stewardship.

Ryan doesn't directly answer my question about how he describes his life, but instead explains how he feels about it. "I started here and I hated it. I *hated* it," he tells me. "It was really low pay and it was a lot of hours, and hard work, and I just hated it. I don't know, over time or what, but now I absolutely love it. There is nothing I could imagine doing different. I love deciding what crops we're going to plant. . . . I love growing the crop now. There's so much more to it than I ever saw as a kid."

I don't doubt Ryan's love for growing crops. I understand the feeling of pride at harvest, the joy of succeeding despite uncontrollable obstacles. But loving to grow the crop is not the same as loving the *work* of growing the crop. Ryan's not the one stooping over rows of lettuce for twelve hours a day. It might be more accurate to say that he loves the job of managing employees, field rotations, machinery repair, and spray schedules required to grow the crop. As such, the work of being a vice president disconnects him from the land and prevents him from taking the role of nurturer that Berry believes is necessary for a farm to be environmentally and socially sustainable. Sadly, I see in Ryan traces of what might be my brother Joshua's future. If our farm continues on its expansion path, Josh will likely oversee the growing of crops the way Ryan does. Josh is already halfway there, given that he's never done several key farm jobs, like harvesting wheat and planting corn. It's not that Dad won't let him; it's that Dad hires contractors to do those jobs because the farm is now too big for two people to complete all the work. I fear that someday Josh will be responsible for overseeing tasks he himself has never done. He'll be a manager.

Like any good manager, though, Ryan cares about the people who work under him, and throughout the day and in subsequent visits I see evidence of his compassion. When we stop to chat with farm employees, Ryan treats them with respect and listens to what they say. He offers praise and credit to employees who stand out. "I actually still don't feel like my job is completely the plants or completely the crop," he explains. "The crop doesn't complain about money. The crop doesn't complain that it's too hot outside today. So my job is still the people that I work with. Even though I love the farming aspect, my job is the people that work for us."

"But I don't want to manage the managers," Ryan adds. By this he means he doesn't want to micromanage or be a "do-what-I-say-and-don't-ask-questions" type of leader. That kind of management can lead to expensive and time-consuming screw-ups. If Ryan tells a tractor driver to plant a block, and the driver knows the block is too wet but says nothing, then the tractor can end up buried all the way to the cab. A backhoe and half a day's work will be needed to excavate it. The tractor drivers know the fields intimately, where they hold moisture, what the soil is like. Ryan has never worked as a tractor driver; he says he bases his decisions on what he sees driving by on the road. He relies on the drivers, and all his employees, to confirm or protest his instructions based on their knowledge. "For me as a supervisor, I want somebody who's not just going to do whatever I say every day," Ryan says. "I want somebody to tell me that what I'm doing is wrong. Otherwise, I can be in control and run the whole thing into the dirt if I wanted to."

Running the whole thing into the dirt is one risk of farming on a large scale. When a farm is big enough to require management positions, dozens of permanent employees, and hundreds of seasonal workers, mistakes happen, since the people running the farm aren't the ones doing the farmwork. I'm reminded of the book *Three Farms* by Mark Kramer, specifically the section called "The Farmerless Farm" where he visits a corporately owned and operated farm in California and explains how red tape, multi-tiered management, and egoism affect food production.[12]

Farm decisions that should have been made immediately—such as when to harvest, apply chemicals, or plant—were instead filtered through layers of management first. The delays in one instance resulted in entire fields of carrots growing too long to fit into standard carrot bags, so the fields were plowed under because the cost of harvesting and processing oversized carrots was higher than destroying an otherwise edible crop. In another case, broken irrigation systems flooded fields in the water-starved San Joaquin Valley. Because the left hand didn't know what the right was doing, fields were sprayed twice accidentally in another instance.

The blunders go on over the pages, all caused by a combination of the farm's huge managerial staff, the tunnel vision of its employees responsible for just one thing and afraid of punishment for overstepping their duties, and the one-size-fits-all approach applied to hundreds of thousands of acres. Similar mistakes happen on large-scale and midsize family farms, too. In a sense, they are just as farmerless. Ryan is much more committed to his work than the people Kramer discusses—it's clear that he cares about the farm and feels a personal responsibility to keep it alive. Still, he's not in the field every day, as he admits. There will always be things his eyes cannot see.

2

The Farm We Grew

How did the modern conventional farm become farmerless?

From the beginning, American farming was about extraction rather than preservation, about settlers civilizing or taming the wilderness instead of working with it. As food writer and chef Dan Barber describes it, "We set out to conquer rather than to adapt—unable, or just unwilling, to adjust our sight to the needs of the new ecology."[1] The New World had a seemingly endless abundance of land, meaning that farmers who exhausted their soils could simply move west—not exactly an incentive to practice good soil management. By the 1820s the nation's farmland bore visible scars: topsoil washed and blown, forests denuded, soil devoid of organic matter.[2]

Still, farms of this era did several things right. The most important thing to note is how diverse they were. Many species of plants and animals called the farm home, which meant each contributed to the overall health of the land. Symbiotic relationships between plant and animal species maintained soil health naturally. Farmers sold their crops and animals in local markets, feeding their families as well as the surrounding community. Farmers were also self-reliant; they saved seeds for the next year's crop, used on-farm fertilizer produced by animals, relied on horses and oxen for muscle power, and controlled weeds and pests through crop diversification and rotation. The sun powered everything, while oil powered nothing.[3] Today we might classify their farming as something close to the "whole-farm approach."

I doubt early farmers would have described their agriculture in such a way, though. The whole-farm approach was really the only approach. Infertile soil couldn't be remedied with synthetic nitrogen like it can today. Diesel power couldn't replace horsepower. Families and communities couldn't be fed with monocultures. What we call mimicking nature today was just farming during this era, much like the distinction we now make between organic and nonorganic food. What we call organic food today was simply food back then. For those who stayed in the East and perfected techniques that mirrored nature, and many did, the whole-farm approach ensured their survival by protecting their resources.

By 1900, though, most of the Great Plains was a patchwork of farms laid out in tidy 160- or 320-acre blocks. Whether the sod was suitable for farming or not, it felt the rip of the plow. Meanwhile, the Native Americans, forced to live on reservations with land so poor that even white settlers rejected it, suffered even more than the prairie.[4] They were destitute, and the buffalo, the main source of survival for the Plains Indians, were practically extinct. As the food market globalized, farmers faced increasing overseas competition and they began to specialize in certain crops and animals to stay in business.[5] Specialization, however, reduced farm resilience, making farmers vulnerable to bankruptcy. If wheat prices plummeted, or disease or hail destroyed the crop, the specialized wheat farmer had no other crops or animals to rely on for income or food. Growing only one or two crops or animals also meant the symbiotic relationships that existed under the whole-farm approach were destroyed and replaced by monocultures. In one analysis of diversification (or lack thereof) in the Corn Belt, researchers noted that "crop and livestock farms became operationally and functionally separate" as farms specialized.[6] In other words, crop growing and livestock raising used to involve intertwined practices, such as the spreading of manure on fields and the feeding of cover crops to animals. That codependence and its resulting ecological benefits disappeared. The researchers go on to note that cattle numbers dropped by 52 percent in the Corn Belt between 1945 and 2000, with hay and oats (i.e., livestock forage) acres declining by 60 percent and 97 percent, respectively, in the same period.[7] The diversified farm died with industrialization.

I didn't realize until I started working for *Tri-State Neighbor* how few farms raise both crops and livestock like my family does. Almost every farm I visited in eastern South Dakota, Minnesota, Iowa, and Nebraska grew only row crops like corn and soybeans. About 80 percent of U.S. corn is grown in rotation with just one other crop, soybeans, with the remaining 20 percent grown in continuous monoculture—corn on corn every year.[8] Cows, sheep, goats, chickens, and horses and the pastures and hayfields that fed them were long gone by the time I got there (in Iowa and Minnesota, some farmers have hogs, but only in confinements). About the only grain farmers who still have livestock are the ones who own land that absolutely can't be farmed, which the USDA terms "wasteland."

Under these conditions, small farms consolidated into larger ones as families went broke or quit farming, a trend that has never reversed and which set the stage for today's megafarms. Specialization also caused farmers to become dependent on off-farm businesses, such as grocery stores, chemical providers, machinery dealerships, and livestock feed suppliers. The typical farm family produced 60 percent of its food in 1900, but by 1920 that percentage dropped to 40.[9] Farm income that was once reinvested in the business went to retailers instead. Most farmers embraced new mechanical tools, such as tractors and grain combines, which boosted production and reduced physical work. Horse and human muscle no longer limited a farm's growth, a development not lost on farmers, who felt they controlled their destiny more than ever before. A 1930 advertisement from the International Harvester Company illustrates the farmer's newfound sense of power and appeals to it to sell the latest tractor.[10] "Be the Master of Mechanical Power," the ad proclaims, with a picture of a man driving the company's new McCormick-Deering tractor underneath. "In these days of mechanical power the man who places his dependence on muscle power is sadly handicapped," the ad continues. "Mechanical power now does the profitable work of the world—both industrial and agricultural. . . . Decide now to make the great step forward. . . . *And resolve to be the master of mechanical power from then on*" (italics from the advertisement).[11]

When I researched today's self-propelled sprayers from John Deere, I found eerily similar rhetoric. "I Run. I Spray. I Conquer. Conquer more acres in less time with the field general in sprayers: the R4045" is the caption next to one of John Deere's latest models.[12] Down the page, I read the lines, "Apply here for total solutions."

Fast forward to 1945, when land conquering took a new chemical turn.[13] At the end of World War II, the U.S. government found itself with ten industrial plants that produced highly explosive nitrate. For agricultural scientists, there was a clear answer to the question of what to do with this nitrate: spread it on America's fields. These bomb-making plants were converted to fertilizer plants, turning the nitrate into pellets and liquids that infused the soil with nitrogen—more tools to master the soil. Synthetic nitrogen fertilizer made natural nutrient recycling unnecessary, so most farmers quickly ditched processes like cover crops and livestock grazing that recharged the soil. Over the decades, a new generation of farmers appeared that knew little or nothing about creating soil fertility through natural processes—they only understood how to apply the right formula of synthetic fertilizer.[14]

Pesticides and insecticides are two more World War II leftovers that became agricultural staples. Other tools developed during that time—such as self-propelled combines, cotton pickers, and bigger tractors—equipped farmers to grow and specialize further. Production exceeded demand and crop prices fell, which pushed more farmers out of business. The farm population dropped to just 15.3 percent of the total population by 1950.[15] Thinking the imbalance in supply and demand was the result of too many farmers, agricultural leaders called for further farm consolidation and mechanization.[16] The 1950s became the era of agribusiness, a term formalized in 1957 by Assistant Secretary of Agriculture John Davis and Ray Goldberg in their book *A Concept of Agribusiness*. They defined agribusiness as "the sum total of all operations involved in the manufacture and distribution of farm supplies; production operations on the farm; and the storage, processing and distribution of farm commodities and items made from them."[17]

Basically, *agribusiness* is a word for the production chain from the farm to the consumer—seeds, machines, and fertilizers, farmers and ranchers, food processors and handlers, marketers who move the commodities and food, and supporters like banks, researchers, and consultants. The message to farmers: control as much of the production chain as you can, mechanize and specialize further, and absorb more farms. Treat the farm strictly as a business. The message turned into a command under the leadership of Earl Butz, U.S. secretary of agriculture from 1971 to 1976. He will be remembered for his five-word edict to farmers, one that haunts agriculture to this day: "Get big or get out." The demanding nature of the statement is unsettling. *Get big or get out.* Absorb your neighbor's farm or lose your own. Buy larger machinery or you'll be plowed under. Make more money however you can or sell out. Think more acres, more technology, more chemicals, and more production, or don't think about farming at all.

Butz directed growers to see themselves as "agribusinessmen," not farmers and certainly not stewards or caretakers.[18] He shaped public policies that forced them to expand and all but ensured they would go bankrupt if they didn't. Butz and his supporters took little pity on farmers who either could not afford to expand or did not believe they should. "As for the farm families who cannot 'get bigger' and therefore have to 'get out,' they are apparently written off as a reasonable, quite ordinary, and altogether bearable expense," wrote Berry in 1977 in reference to Butz's attitude.[19] If you couldn't get big, then you deserved to fail, the thinking went. "Get big or get out" was not advice, it was a threat.

Besides "get big or get out," another of Butz's infamous ideas was that farmers should plant "fencerow to fencerow." A fencerow is the strip of uncultivated land between fields, which farmers implemented after the Dust Bowl to help control erosion. Farming fencerow to fencerow meant eliminating all barriers to production, even necessary ones like fencerows that conserved soil. Physically and symbolically, his "fencerow to fencerow" policies undid much of the conservation progress that had been made after the Dust Bowl—terraces, contour plowing, no-till agriculture, crop rotation, riparian buffer zones, and lines of trees known as "shelterbelts." By calling on farmers to abandon conservation techniques and create megafields instead, Butz asked them to return to the mental and physical landscape of the pre-1930s. Butz also dismantled the government's production restriction programs and grain reserves intended to balance supply and demand, which encouraged all-out production.[20]

Farmers responded with vigor to the "get big or get out" order, plowing up what little native prairie remained, expanding fields, dozing in tree lines, and ripping up fencerows. Ranchers crammed their ranges with livestock. They borrowed heavily, a mistake that would cost many their farms a decade later during the 1980s farm crisis.[21] It's surprising how little the leaders of the 1970s remembered or cared about the nation's worst environmental and social disaster, the Dust Bowl. After all, it was only forty years in the rearview mirror. Butz was in his twenties during the Dust Bowl, certainly old enough to internalize its full horror. Yet his concept of "full production" shifted agriculture back to Dust Bowl–era thinking, and we're still operating under these principles today, even as climate change brings the threat of worsening droughts and floods.

My dad's father, Marlo, was one of those who answered Butz's call. He built grain bins and machine sheds, bought combines, tractors, and pickup trucks, leased and bought thousands of acres of land, enlarged his cattle herd, and was "pivotal," as my dad puts it, in bringing a grain elevator to the nearby town of Bison so that farmers could produce even more grain.[22] He also built a feedlot for fattening calves for CAFOs (these had appeared nationally in the 1950s and '60s as corn prices lowered).[23] His expansions continued until his death in 1997, all involving heavy debt that my parents and relatives were forced to pay off when he died. They had to sell land to do so. If anyone eagerly threw his hat into the agribusiness ring, it was Marlo. He was the person who taught my father how to farm—well, he taught my dad one version of farming—and it's easy to see Marlo's agribusiness ideals in my father's current operation.

My dad well remembers these farm expansions of the '70s, while he was in his late teens and early twenties. He recalls hearing "fencerow to fencerow" on TV and reading the phrase in newspapers. "In the '60s, it seemed like guys were still running the old tractors, the old WD's and the old M's," he told me over the phone once.[24] I'd always known my grandparents were considered big farmers—my dad had the nickname "money bags" in high school on account of the farm's major purchases, the debt notwithstanding—but I was curious to see how my dad remembered it. "Then in the '70s, stuff started getting bigger. We'd think it was small today, but it was big then. It ended up haunting us, overproduction with no market."

Prices were high, the government demanded more production, and the agribusiness sector was churning out amazing new equipment, agrochemicals, and hybrid seeds. The high times of the '70s convinced many that Butz's maxim of "get big or get out" had worked. Hindsight, though, proves otherwise, as my dad correctly points out. During the farm crisis of the 1980s, thousands of farmers and ranchers went bankrupt as commodity prices dropped to 1960s levels and below.[25] For example, my mother's parents, cattle ranchers operating twenty miles away from Marlo's farm, declared bankruptcy in 1982, a direct result of over-borrowing to expand their herd.

One of the most troubling effects of the farm crisis, however, was the weeding out of midsize farmers and ranchers like my grandparents. As agricultural writer R. Douglas Hurt argues, farm policy in the 1980s "worked to the advantage of large-scale farmers, enabling them to purchase more land and equipment and thereby farm more extensively and efficiently while forcing small-scale, noncompetitive, or unprofitable farmers from the land."[26] In those years midsize farms broke even, if that. Small farms were practically impossible to sustain without off-farm income. As

Hurt rather glumly concludes, "Given these problems, one should not be surprised at the number of people who left the farms. Rather, one should be amazed at the number who remained."

Why couldn't farmers like my grandparents—and today people like my father, the leaders of Roth Farms, and other conventional farmers—see through the "get big or get out" lie? Wendell Berry claims people bought into industrial agriculture mainly because it promised a better future, a promise that turned out to be false. He also points out that the adoption of conventional agriculture was a gradual process, not an instant transformation, meaning its consequences weren't immediately clear, but became so many years later. As he observed: "A great deal of the strain of the industrial revolution has been borne by farmers, and so it has been fairly easy to secure their allegiance to the future, when more industrialization will supposedly bring a better farm economy. The industrialization of farming as we now have it is not something that farmers would have bought all in a piece; as a group they have been too traditional or conservative for that. Industrialization has been sold to farmers in stages, one implement at a time."[27] One implement at a time: buy this new machine, stop growing this crop, specialize in this, quit using that natural process. Each decision represented a supposed improvement, and given the American obsession with bettering ourselves, farmers understandably embraced what they were told was progress.

After one too many "improvements," though, farmers woke up to find themselves trapped, too invested in industrial agriculture and in debt to turn back but barely clearing expenses. John Ikerd calls this spiral of events the treadmill of industrialization, which works like this: the promise of profit baits farmers to expand, more technology and acres or animals increase profits temporarily, then prices fall because of overproduction since everyone else also gets bigger.[28] A price crisis ensues, farmers are pushed out of business, and those left on the treadmill have to run faster by getting bigger and producing even more. Instead of working together, farmers compete, with the "winners" eventually closing in on the "losers" and swallowing their farms. Writer Eric Schlosser calls this thinking the fallacy of composition, or "a mistaken belief that what seems good for an individual will still be good when others do the same thing."[29] The only way to make money in the industrial farm model is to boost yields, push down expenses, and grow more of your specialized crop. Farmers accomplish this by applying more fertilizer and agrochemicals, acquiring more land, forgoing crop rotations, farming up conservation areas, and planting GM seeds. They achieve their goal, more yield, but so has everyone else, which depresses prices further, which means farmers need to produce ever more to stay in business. As bizarre as it is, though, farmers insist on gauging their success in terms of yield, which goes up even as they go bankrupt.[30]

This is the farm we've built: one that is productive to the point of destruction, heavily industrial, and controlled by retailers and processors—where living is stressful, stewardship is gone, and the tools have undone their operators. We can see this by taking a hard look at the state of agriculture in America right now. According to the 2012 Census of Agriculture, the United States has 2.11 million farms, about 100,000 less than in 2002.[31] A very small number of these farms holds the majority of the land and produces most of our commodities and food crops. About 1.7 million farms—more than 80 percent—did less than $100,000 in farm sales, together representing 5 percent of the total sales. That means about 20 percent of the farms produce 95 percent of the commodities and food. What we have is a top-heavy agricultural system, with a small number of very large farms providing most of our food. The middle is all but gone.[32]

The midsize farmers who are left might argue that they are separate from this top-heavy agribusiness system; that they are not who we mean when we talk about industrial farming. But every producer who engages in conventional farming and ranching is playing by the rules of agribusiness. It's unavoidable now. Most midsize and small farmers are planting monocultures, spraying herbicides and pesticides, and sending their cattle to CAFOs. Unless those farms are organic or sustainably managed, they are industrial, no matter their size.

Judging from our agriculture, the Dust Bowl has receded from America's collective memory, an extraordinary feat of nationwide amnesia. Relying on Dust Bowl–era practices poses greater environmental risk now than at any time in human history. Climate change is causing severe droughts, floods, and other erratic weather, and our farm and ranch land is poorly equipped to handle the changes. Yet the industrial model of agriculture does not take externalities like land or water into account as it calculates efficiency and cost. Our model only considers input versus output, and yield per acre per farmer. This is why people often say American farms are the most efficient in the world—and in a narrow input/output sense they are highly efficient—but this efficiency is hollow in almost every other sense.

That is the real state of agriculture: blind. Blind to the land, human health, rural communities, and the farm family. Blind to the future, which depends on preserving resources. Blind to everything but profit.

3
The Growth of Roth Farms

It's a late May morning after Memorial Day, hot, humid, and sunny, with wide and puffy clouds passing overhead. On the northern horizon, a bank of rain clouds mushrooms into the sky, a storm that will break upon us later. The ground is wet from more than an inch of rain the night before. Rice grows in the fields along the road, flooded and green. Ryan and I are "riding cane," which means driving by in the pickup and scanning for diseases like orange rust or brown rust. Alligators eye us in the canals as we pass.

Ryan identifies the different varieties of cane, some that yield higher tonnage, some with higher sucrose levels, others that grow faster and close quicker to choke out weeds. Some cane has a thin blanket of rust on the leaves, as he slows down to show me, while other cane is bright green, shining with health. With so many farmers growing sugarcane in the area—especially the sugar corporations like Florida Crystals and U.S. Sugar—being big is the only option here. "You're gonna have to be a bigger producer," Ryan tells me as we pass walls of green stalks. "The smaller guys are gonna get pushed out or get bigger." And get bigger is exactly what Roth Farms did.

One way the farm expanded was by adding turfgrass. Turfgrass production involves heavy watering and frequent mowing followed on some farms by a vacuuming of the clippings. It's tough to keep up with an industrial-sized lawn, but with new houses, golf courses, and athletic fields springing up across South Florida, the Roths enjoyed high demand for their turfgrass for years. When the housing market collapsed during the 2008–9 recession, though, so did the sod market. Sod plummeted from fifteen cents a foot to seven cents, Ryan tells me, and Roth Farms went from harvesting one hundred truckloads of sod a day to just fifteen.

The Roths were forced to shut down the harvesters that had been motoring over the fields for decades—but that wasn't entirely bad in Ryan's estimation. "Sod was probably the most profitable crop we were growing," he says. "But when you harvest grass, you are sending muck out of the farm."[1] That's one major drawback of growing turfgrass: at harvest, machines scrape it off the field, keeping the roots intact but slicing away a thin layer of dirt. The machines either spool the grass into rolls or cut it into slabs. Either way, the process strips away soil. "To me that wasn't good for the long-term ability to farm, because I want to be here thirty to forty years," Ryan says. "If they are selling with the sod an inch of muck every time they harvest, that wasn't something I was happy about. I wasn't glad that the housing market fell apart, I wasn't glad that it cost us a bunch of money, but I'm glad that we stopped growing grass."

Ryan correctly realizes that something good came out of the crash: the farm became more sustainable. If the decision had been his alone, then the farm would have eliminated turfgrass sooner. "It took us two years longer to get out of the sod business than it should have because it was emotional," Ryan says. "We thought, we have people who are sod harvesters and they'll have to quit. And we've been doing this for years, and we have a separate company called Roth Sod and this is *what we do*. Other farms said, 'Psh, there's no money in sod, let's get something else planted here.'"

I commend the genuine concern about workers losing their jobs, but Ryan's explanation mostly reveals that acting quickly in response to changing conditions isn't easy for farms as large as Roth's. Getting out of sod wasn't a matter of simply planting something else; it meant fundamentally changing the farm's business model. This is why industrial farmers feel like they can't diversify or stop growing the crops they've specialized in—

they've sunk capital into the equipment and infrastructure needed to grow a handful of crops in a big way. Growing a new crop means starting a new business, which most can't afford or don't want to do. In contrast, midsize and small farms tend to be more flexible and resilient in the face of change, and that is a good trait in a volatile world made even more so by climate change. Midsize farms in particular would be capable of responding to consumer and government demands for regenerative practices—they have more access to financial resources than small farms—and they would also have a bigger impact on the food supply because of their size and output.

While he says he is happy overall, I think Ryan is chaffing a bit under the yoke of conventional agriculture, though he might not put it that way. "There are certain crops that are just completely and totally uninteresting to me," he tells me. "Radishes are not interesting. Do we grow them?" He grimaces and nods. Why aren't radishes interesting? I ask later. There are two main types of radishes, Ryan explains, open-pollinated (OP) radishes and hybrid radishes. In the U.S., growers can buy just one OP seed variety called red silk—the bright, cherry-red kind found in most stores. Given that radishes aren't very popular in the U.S., Ryan doesn't expect researchers to develop new OP radishes anytime soon. He's tried growing hybrids, but the seeds are four times more expensive, and it's already tough to make money on radishes. Plus, consumers don't want hybrids, he says. He's tried different colors and sizes, such as watermelon radishes that have red flesh, green skin, and a somewhat sweet flavor, but he couldn't sell them. So every year, Ryan plants red silk radishes—and that's it, which is why radishes aren't a very interesting crop to him. "I like looking at new varieties and looking at how different things respond to weather and yield differently, but there's just nothing being done there," he says about the radish business.

As Ryan tells me all this, my mind jumps back to the black spot issue. One way to combat black spot is to plant on "fresh" ground, or fields that haven't had radishes for a while. But another way would be to grow different varieties that don't easily succumb to the disease or resist it altogether.

Switching varieties every so often would also disrupt the pest cycle. Ryan doesn't have the option of using these natural processes, though—and it's a result of the agribusiness push for standardization, which eliminated all OP radish varieties but one. Years of standardization redefined the consumer's idea of a radish, which is now limited to small, red, and round. Because agribusiness research money goes toward profitable crops like corn, cotton, and soybeans, radish producers are stuck. If anything ever happens to the red silk variety—a bug or disease makes it impossible to grow any longer, let's say—then there's no backup option. In the case of radishes, we've standardized almost to the point of disappearance.

Ryan has more options with sugarcane, one reason he finds it more interesting to grow. "You can play around a lot planting different varieties, trying to maximize production on certain farms," he says. He enjoys the intellectual work—proof that when farming becomes too automated, the farmer can get bored, but when it's hands-on, the farmer is more engaged. Sugarcane is a four-year crop; the same plants are harvested once a year for four years. The harvesting machines leave one or two inches of the plant behind and it grows back. At first I am amazed by this fact, but then I remember that sugarcane, a member of the Poaceae or grass family, is really just oversized, sugar-heavy grass.[2] Of course it grows back, just like a lawn. Definitely interesting. No matter how interesting a crop is, though, it has to be profitable. Ryan sees little room for sentimentality in farming. "You do have to love it, but you do have to be able to react. You can't stay in a crop just because you love it. You can't love your crop. You can love your job, but you can't love your crop. You have to be able to say, 'That doesn't make me money anymore. Let's do something else.'"

You can't love your crop. Ryan's words might sound harsh, but he is right. Farmers and ranchers, even regenerative and organic ones, can't raise crops and livestock that consumers don't want; if they do, they'll go bankrupt. But if the rise in organic food sales over the last decade is any indication of the future, then there will come a day when most consumers don't want the crops Roth Farms and other industrial operations produce. Conventional farmers will have to look at their fields and say, "That doesn't make me money anymore. Let's do something else."

The most significant "get big or get out" move at Roth Farms came in 2005: the construction of a commercial produce packinghouse.

The idea was to vertically integrate, to control more of the supply chain and therefore a bigger portion of the profits. Roth Farms was selling vegetables wholesale at market price to commercial packers, who packaged and washed the vegetables, then turned around and sold them at higher prices. If Roth Farms owned a packinghouse, the board thought, then there would be no middleman between them and the grocery store. Other farms in the area and across the country had already made such a move. Ryan explains the economics behind the decision: "We grow vegetables and we pay other people $1.25 to cool every box that we grow. It probably costs them somewhere in the eighty- to ninety-cents-per-box range to cool that product, so they were making thirty to forty cents on every box. We grow the box, and they get thirty to forty cents profit. Not revenue, *profit*. There is no guarantee of thirty or forty cents revenue or profit on the farm. It may be a loss. The packinghouse is almost guaranteed money. So the idea was that we could build a packinghouse, and if we grow more product we can make more money. Also, you look around Belle Glade, there are farmers that go broke," he continues. "Packinghouses never go out of business. Something goes right at the packinghouse. It's almost guaranteed money." The second time he says this phrase, he emphasizes *almost* instead of guaranteed. "There's not as much risk as farms have. Farms have a lot of risk."

The desire to mitigate risk is awfully appealing to farmers. When people say farming is like gambling, they aren't far from the truth. Inputs like herbicides, fertilizers, and GM seeds appear to reduce risk, but a farm's success is also intimately connected to the weather—which is becoming more erratic because of climate change—and other uncontrollable factors such as commodity prices and consumer taste. But acquiring another link in the farm to the consumer chain turned out to be much different than the Roths expected, largely because of the 2008–9 recession. "We built it with the idea that even if vegetables were bad, sod production was so good that it would pay all the bills and it would be no problem," Ryan says. "Then the housing market crashed and we almost went out of business."

Roth Farms was in debt to the tune of $11 million and their main revenue stream had dried up. They received offers of $6 million and $7 million for the packinghouse, which they understandably rejected. They elected to keep it and sell land to cover the payments instead; they reasoned that if they took a low offer, they still would have to sell farmland to pay it off, so why not keep the asset and hope its value increased. Still, the situation remained grim. "I was trying to decide what I was going to do if the farm went out of business," Ryan says. "I guess I was going to go back to school and be a lawyer or something, I don't know." The Roths didn't intend to over-leverage quite so much. Two investors backed out at the last minute, leaving them and one other farm to foot the bill. "What we should have done at that very moment was cut the size of the thing in half where we could spend less money, have it just be big enough to run our packages," Ryan says. "Looking back on it, I wish we hadn't built it. We had a really good farm that we had to sell in order to keep it afloat. A two-hundred-acre farm, probably the best farmland we had, we had to sell."

The packinghouse stands on Belle Glade's northern edge, near the municipal airport. Next door is TKM Farms, whose packinghouse supplies lettuce to Taco Bell and Kentucky Fried Chicken. The Roth packinghouse isn't Belle Glade's largest, but it's one of the newest and coldest, as Ryan describes it, meaning the building's fancy cooling mechanisms and modern airtight design can maintain low temperatures better than older, less efficient packinghouses. The parking area is nearly empty on the May morning I visit, with only one forklift buzzing pallets of sweet corn to and fro. The packinghouse is tall and gray and studded with loading docks, like an oversized Wal-Mart without the sliding doors. Rain pours off the roof in waterfalls; the gutters can't keep up. A shroud of mist hovers around the building, named "Ray's Heritage" after Ryan's grandfather, Ray Roth, who founded the farm in 1948.

Huge boxes of sweet corn still in the husks wait to be carried inside on pallets. Ryan says corn surrounded the packinghouse from morning until night in the two weeks before I arrived, and inside it was stacked clear to the ceiling. Trucks lined up down the road to pick it up. "Today it's a ghost town," Ryan laughs. Such is the volatile nature of supply and demand. Springtime is the main sweet corn season in Florida, and growers time their harvests to meet the big sweet corn rush over Memorial Day weekend (Memorial Day was Monday; I came the following Thursday). One-third of America's sweet corn originates in the Everglades Agricultural Area.[3] "Belle Glade is the sweet corn capital of the world," Ryan says. "There are two million crates pulled in a two-week period out of this area. If you eat sweet corn on Memorial Day somewhere inside the United States of America, the likelihood is it came from here."

He says this with pride, and it is an astounding fact. I think of the boxes rattling inside a truck from here to Bison, over two thousand miles, and ending up in the town's tiny grocery store, where Josh rummages through the box for a dozen big ears to eat on Memorial Day. And the corn doesn't stop there—it keeps going to Texas, California, New York, Washington, Maine. I imagine Belle Glade as a heart pumping corn down the highway arteries. One section of drained Everglades, a tiny area when compared to the whole nation, can supply almost all of America with sweet corn for a few weeks. Just a handful of farmers produce this corn, and I'm sure Ryan knows most of them. It's a testament to our top-heavy, shipping-dominated, increasingly fragile food system.

He shows me the radish line and the green bean line and then two "cold boxes," or giant warehouse-sized rooms that store vegetables until they leave on trucks. Overhead fluorescent lights illuminate corn boxes stacked almost to the ceiling, more ears than I've ever seen. Everything shines—handrails, stairways, even the walls. The cold boxes, and the whole packinghouse, have high ceilings so that warm air floats up, away from the vegetables. "There are no food safety problems in this building whatsoever," Ryan says proudly. "It's clean, it's beautiful, it's perfect."

The packinghouse may be top of the line, but it's a liability. Roth Farms is now stuck growing green beans and radishes because they have specialized packing lines for those two crops. "Green beans have been terrible the last few years. We can't make any money with green beans right now," he says. Still, they have to keep growing enough beans to keep the lines running, or buy beans from other farmers to supplement their own. The radish line is the result of the farm's emotional attachment to growing radishes. It was a tradition, something Roth Farms had done since Ray's time. Investing in those lines was the equivalent of Roth Farms saying, "We love radishes and green beans"—and we know the rule about loving your crop. You can't. "We spent a lot of money to build that pretty radish line. Now we have to grow radishes for the next fifteen years until we pay it off," Ryan says. "I wouldn't have done it. I wouldn't have built that radish line. I would have quit growing radishes."

Leafy vegetables also end up at the packinghouse, but unlike radishes and green beans the leaf gets sorted and packed in the field. The goal is to harvest, cool, and ship the highly perishable lettuce all in thirty-six hours. Ryan explains that packinghouses used to hold lettuce for seven to ten days before sending it off, a practice that didn't contribute to freshness—or to food safety. Lettuce and other leafy vegetables are extremely susceptible to salmonella, *E. coli*, and Norovirus. A 2009 report, ominously titled "The Ten Riskiest Foods Regulated by the U.S. Food and Drug Administration," analyzed reported foodborne outbreaks between 1990 and 2006. The researchers found that leafy greens were the nation's most contaminated food, accounting for 24 percent of total outbreaks.[4] Keeping leafy greens clean and safe isn't easy. Animals, manure, water, or improper handling can contaminate them. Pathogens tend to linger and grow until consumers eat them. Postharvest treatments such as chlorine washes reduce contamination, but they aren't perfect—bacteria can live in the washing systems, which can contaminate batch after batch of prewashed bagged greens.[5]

Roth Farms has never had issues with contaminated greens or other vegetables. Ryan attributes this to their careful compliance with food safety rules, such as providing hand-washing stations and gloves and preventing workers from eating in the fields. The farm keeps a food safety manager on staff year-round, a move that goes beyond federal regulations. He also points out that the EAA's topography—flat with very little runoff—and lack of livestock help ensure safe food. Still, the potential for contamination is always there. When Ryan's produce leaves on a truck, for example, he cannot control how it is handled. I ask Ryan what it's like to carry the burden of food safety on his shoulders—the guilt if someone were to be sickened or, worse, die, the danger of lawsuits that could sink the farm, the shame of putting out "dirty" produce. "It's extremely terrifying because I'm not in control of my own product before it gets to the consumer," he says. "The likelihood, if somebody gets sick, is that the first person they are going to talk about is the farmer. [They will say] the product coming out of the field is bad. But there are so many different inputs, so many different people touching the crop, the product, before it makes it to the person's plate. It does scare you to know you're sending food out. I'm not growing something that gets cooked. It's all raw. There are a lot of things that make you uncomfortable. I feel a lot better about sweet corn and green beans because people eat them cooked."

The long and uncertain chain from farmer to consumer and the lack of control the farmer has over the product once it leaves the farm—these are the consequences of the industrialized food system. Farmers live in fear of contamination that will be blamed on the farm even if it didn't originate there. And with dozens or, for the super-sized farms, hundreds of workers handling produce, on-farm infection is a real concern. Meanwhile, the consumer lives in fear of foodborne pathogens that can sicken or kill, pathogens that spread quickly and invisibly through massive food processing facilities and distribution networks. *E. coli* that originates in a lettuce processing plant in one city can easily appear across entire regions or the whole country.

But Ryan can't market his produce locally or even regionally because buyers in the Southeast don't pay enough, he says. Florida farms fight for a share of a commercial vegetable market dominated by California and, increasingly, foreign imports—and they have a hard time competing. Southeast buyers reason that they shouldn't have to pay as much for Florida products because that food doesn't have to be shipped as far, so they offer less. Ryan's produce ends up with the highest bidder, which is usually thousands of miles away. "Say lettuce is ten dollars a box. I'm making very small margins at ten dollars," Ryan says. "The Southeast buyers want to pay seven or eight dollars. Up in New York, they'll pay ten dollars. I would much rather put it on the truck and get it up there to them, but I'm losing some control. What would be great is if local people down here would spend a little bit more money on their vegetables."

Getting bigger, it seems, hasn't translated into higher profits, less risk, or more security for Ryan and his family. Far from being the silver bullet Butz promised, getting big has shackled them to a small number of low-profit crops, saddled them with debt, separated them from the consumer, and removed them from the land and actual work of farming. This is what family farming looks like all over America—but despite the farm's size, Ryan insists that his is a *small* family farm, as similarly sized, conventional grain or dairy farmers in other states would likely argue. "In our area, we're a small producer," Ryan says. "We're not a big farm. Here? Here, we're dealing with the likes of 150,000-acre sugarcane farms. We're small. We're a real small farm compared to that."

I take Ryan's point. Roth Farms is definitely modest compared to its neighbors. Maybe size, like beauty, is in the eye of the beholder, or best understood in comparison to the surrounding farms. In my eyes, Roth Farms is big, not because of its acreage but because of how it operates. With a commercial packinghouse, corporate-like structure, and input-heavy production model, Roth Farms plays by the "get big or get out" rules. If the conventional American farm can't afford to stay small, Ryan tells me, then it should try to act small in as many ways as possible. "If a small farm in acreage, size, and personnel won't survive because of the new agriculture, then strong family farms have to figure out a way to remain in business and get bigger, and continue to act like family farms," he says.

Ryan's dad, for example, does farm tours, bringing grade-school kids out to the fields. Ryan names workers who've been on the farm for thirty, twenty-five, and fifteen years, testaments to the good working conditions. He tells me about an end-of-the-season party his dad and uncle threw during one of the farm's toughest post-packinghouse years. The Roths gave the workers a bonus, cooked food, and made Roth Farms T-shirts. The radish-line ladies immediately went to the bathroom and changed into their new shirts. "They came out and said they wanted a picture. We didn't tell them this was what we wanted to do. They said they wanted a picture with me, my dad, and my uncle," Ryan says. "We were about to go out of business. They didn't leave and go find other jobs that were more secure. They stayed with us. We took care of them."

The Roths are good people who care about their employees and act small when they can. But acting small in some ways is not enough to counteract the problems of acting big overall. In his book *Let Them Eat Junk*, Robert Albritton argues that even small, family-owned farms become pawns of agribusiness when they adopt the conventional model, because farming conventionally means using the inputs provided by agribusiness corporations and growing crops and livestock according to agribusiness standards.[6] Conventional farmers become reliant on inputs—herbicides to kill weeds, fuel to run tractors, GM seeds to resist sprays. Meanwhile, food processors insist on standardization. Vegetables must be a certain size, weight, color, and variety. Wheat needs to have a certain protein percentage. Specific standards apply to chickens, hogs, cattle, grain, milk, eggs, virtually every product intended for the mass market. If a farmer's output (crop or livestock) does not meet the standard, then it cannot be sold. Ryan being forced to grow red silk radishes is exactly what Albritton is talking about.

The conventional farmer, no matter how big or small, must comply with agribusiness specifications or lose money—comply or die. These specifications predetermine the farming process, driving farmers to adopt a business model that helps them meet these standards and ditch crops or practices that don't. Regenerative practices are usually the first to go since their benefits cannot be calculated on a profit-and-loss sheet. It's not just food processors setting the standards, it's machinery dealers, chemical and seed companies, feed sellers, crop insurance companies, anyone who represents a link on the agribusiness chain between farm and consumer. Farmers, including Ryan, tend to see themselves as responding to consumer demands, and that is how they explain their choices. But compared to the power of agribusiness corporations—in shaping government policy, funding university research, influencing consumer views, and controlling the food chain from farm to grocery store—the consumer shapes very little of the market, while agribusiness forces shape most of it. Farmers aren't growing for the consumer, but for corporations.

4
The Farm Town

People say Belle Glade is known for three things: sugarcane, prisons, and National Football League players.[1] The town has other distinctions as well. In 1985 it became famous for having the highest AIDS infection rate in the nation, higher than New York City, ground zero of the epidemic at the time.[2] In 2003 the town made national news for having the nation's second highest violent crime rate per ten thousand residents.[3] Poverty is long entrenched in Belle Glade, as shown in the 1960 documentary *Harvest of Shame* that depicted the desperate lives of the town's farm families: starving children, rat-infested homes, and thin, tired, typically black mothers, fathers, and teens crowding into trucks bound for vegetable fields.

Things haven't improved much since the film appeared. Unemployment in Belle Glade hovers around 40 percent today, and 33 percent of the town's people live below the federal poverty line.[4] Decent-paying jobs just don't exist, except at the correctional facilities. People used to work as sugarcane cutters until the area's sugar corporations—collectively known as "Big Sugar"—replaced human workers with machines in the early 1990s.[5] What really hurt the economy, though, was Big Sugar's buy-up of the region's vegetable farms, which drastically cut the number of agricultural jobs and drove away businesses that depended on local farmers and their families. Almost everyone with means to leave Belle Glade has done so, including what family farmers are left. "Even me, I'm proud to work out here, but I can't live out here," says Ryan, who lives in a town called Loxahatchee, twenty-five miles away from the farm.

Belle Glade is one of thousands of small, rural communities in America that used to rely on small and midsize farms for economic support. That sector is shrinking nationally, and it's all but disappeared around Belle Glade. When farms get bigger, local economies suffer. One reason is that big farmers rarely do business locally because they can find cheaper goods elsewhere. They buy in bulk on the national or international market. It's not just about economies, though, it's about communities. As John Ikerd writes, "A rural community is far more than a rural economy. It takes people to fill the church pews and school desks, to serve on town councils, to justify investments in health care and other social services, to do the things that make a community. As farms have grown larger and fewer, rural communities have lost people—human and social resources—and many rural communities have withered and died."[6] In rural towns, the social cost of industrial farming is on full display. I see it in my hometown of Bison, a lonely farming community with a population of 338, half of what it used to be decades ago. What few businesses are left struggle to stay open, with the exception of the grain elevator and stores that sell fertilizer, seed, fuel, and agrochemicals. If Bison wasn't the county seat, then the courthouse and its jobs wouldn't exist for the many farm wives forced to take employment in town. More people are leaving Bison than staying and raising families. Myself included.

Who can blame them for leaving? Not that I'm trying to justify my flight, but Bison's heyday is long past. The nearest movie theater is 50 miles away, the nearest commercial airport 150. There's one restaurant, one tiny grocery store, one bar (well, if you count the bar inside the one convenience store, two), and zero stoplights. One of my sisters graduated high school with just four other students. It's not that fun to live in Bison anymore; people often feel isolated and bored. Young people are choosing city life instead of taking over the farms they were raised on, which allows the area's biggest farmers to swallow up more land. If nothing changes, then Bison could die. And it is just one such community among hundreds of thousands on the verge of collapse, not to mention the thousands more that already have.

For the time being, Belle Glade still offers some agricultural jobs, such as planting, cultivating, spraying, and harvesting vegetables. When I visit Roth Farms on a windy day in November, I watch members of a harvesting crew stoop, sever iceberg lettuce heads with their knives, slice away the limp outside leaves in one or two quick motions to leave just enough "wrapper leaves," and lay the heads on the bed. Other men pack the heads into boxes, and still others, the loaders, lift the boxes onto moving flatbed trucks. A box can weigh between forty-five and sixty pounds depending on the crop packed inside, and a crew pulls roughly two thousand boxes of leaf in a day.

Ryan stresses that field workers are paid good, but not great, wages for this tough physical work. He says leaf harvesters earn roughly $600 to $800 gross a week on his farm, loaders around $1,000 (the number of days crews work in a week varies according to field conditions; same with the number of hours worked per day). Ryan's lettuce crews earn more than Florida's tomato pickers, who also labor under the piece system, earning $70 per day on a good day. But for all field workers, the piece system is unfair. If rain or dew keeps them out of the field for an hour or two, they stand around, earning nothing, most unable to go home because they don't own a car. Some days they earn nothing, and most receive no benefits. There's also no overtime pay. If a storm is coming and a farmer needs the crew to work over the weekend in order to save the crop, he or she is not required to pay them more, even if the workers have already put in forty hours that week.[7]

As we watch the loaders, Ryan comments that he can't believe they can work so hard day in and day out. A loader might lift 120,000 pounds in a single day, he says. "All these are jobs Americans don't want to do," he says, gesturing across the field. "But that job specifically is one." I see what he means: every person in the field is of Hispanic descent except Ryan

How can we change this?

Who are these people?

and me, and he tells me later that most are migrant workers here on work visas. The crews work about six months of the year for Roth Farms, then move up the East Coast for the summer. I feel ashamed walking through the field, taking notes about the sweat on their brows and the way they hunch over the rows. The pain they must feel in their knees and backs, the cuts they must endure on their hands and arms from swinging the knife. All so people can eat a one-dollar burger with a couple of lettuce leaves from a fast-food drive-through.

Conventional producers can't charge more for their crops to cover rising input costs, but they can control what they pay for labor, which is why farmworkers are some of the nation's lowest-paid laborers. Only dishwashers earn less. Most farmworkers live below the poverty line, on minimum wage and with virtually no benefits. Less than one-tenth receive employer-paid health insurance, and just 10 percent receive paid holidays or vacation time.[8] Factor in the extremely hard work, and most Americans, like the residents of Belle Glade for example, cannot afford to take farm jobs even if they wanted them—but desperate undocumented immigrants or impoverished migrants on work visas often will. The situation is similar in California, where Mexican workers pick grapes and strawberries in withering heat. On the Great Plains, wheat harvesting companies sponsor cheap workers from South Africa; these are the people who've harvested my father's wheat in the past. In the dairies of Wisconsin and Minnesota, Central American workers draw milk from thousands of Holsteins each day for minimum wage. Meanwhile, people from the nearby farm towns cannot find work.

Farm labor in the U.S. has always been a job of the lowest socioeconomic class, one reason we don't value such labor in our society. In the South, slaves worked cotton, sugarcane, and tobacco plantations. After the Civil War, poor, usually black sharecroppers lived as serfs under wealthy landowners. In Texas, during the late 1880s, cattlemen hired expendable labor, called cowboys, to drive cattle thousands of miles across dangerous terrain to market. Though the Chicago meatpacking plants of the early

1900s were not farms, they, too, preyed on poor immigrant workers from eastern Europe to slaughter and process livestock, paying the workers next to nothing and providing no benefits when they were injured or killed. In California during the Depression, farmers who had lost their farms, derisively called "Okies" because many were from Oklahoma and surrounding states, picked fruit and vegetables for pennies a day.[9]

Putting impoverished foreign workers in U.S. fields became institutionalized during World War II. In 1942, when the West Coast experienced a wartime labor shortage, the federal government invited Mexican workers to American farms under the *bracero* program. The government pledged transportation assistance, living expenses, and a return trip to Mexico when their work agreement ended. Growers saw an opportunity to hire a controllable workforce that they could pay a fraction of what they did American workers. West Coast growers became so addicted to cheap Mexican labor that after the war they persuaded Congress to prolong the *bracero* program, which lasted until 1964. The arrangement ultimately pushed domestic workers out of the fields and caused devastating poverty among the Mexican field hands, who lacked the ability to protest their situation.[10]

Ryan would like to see immigration reform, specifically a law that would allow foreign workers to return home during the off-season. They long to go home, he says, but if they do they likely won't be able to come back. "Getting back is going to be such a huge problem that they just stay, and they may stay for fifteen or twenty years, making as much money as they can, sending money home, trying to survive here, so they can get back home. If you did set up a work permit program where people could come and say, 'I have a job at Roth Farms' or 'I have a job at whatever'—it may not be Florida, it may be New York or California or whatever—but 'I have a job, it's a six-month-a-year job. They want me to come work this job and those other six months I will sign documentation that I'll go back.' Because that's what a lot of these guys want to do. Some of them want to work here in the wintertime and go home in the summertime, some want to work up in New York or New Jersey or wherever there are other farms in the summertime where they grow vegetables, and go home in the wintertime. They can't do that. It's really sad to hear some people say, 'My dad died. And I can't go home for the funeral.'"

Life for migrant workers on visas is often dismal—but at least they have a choice to stay or leave. When writer Barry Estabrook published *Tomatoland*, his exposé of Florida's tomato farms, in 2011, police had rescued more than a thousand people from Florida farms who were being held against their will and forced to work—and Estabrook says that number is "only the tip of the iceberg" because most cases go unreported.[11] In South Florida's agricultural world, he writes, "slavery is tolerated, or at best ignored." He describes human traffickers promising men and women jobs in America, but these people arrive in Florida only to discover they've been sold into farm crews, where guards lock them in shacks at night and beat them if they attempt an escape. These workers often have no idea where they are, can't speak English, and are in the U.S. illegally, which makes them hesitant to approach police if they do escape. The bodies of murdered field workers routinely show up in the region's rivers and canals, even now.

The human trafficking situation has improved since 2011, but the appalling conditions in which legal, non-enslaved fieldworkers live and labor have not, especially in terms of housing.[12] Workers usually live in overcrowded, decrepit trailers and apartments, sleeping on floors but paying high rents to rural slumlords. They can barely afford to feed themselves, so they go to soup kitchens and food pantries as if they were homeless. As researchers have noted, every day on the job is dangerous for fieldworkers: "The rate of death due to heat stress for farmworkers is twenty times greater than for the general population. . . . Fatality and injury rates for farmwork rank second in the nation, second only to coal mining. The U.S. Environmental Protection Agency (EPA) estimates that U.S. agricultural workers experience 10,000 to 20,000 acute pesticide-related illnesses each year, though they also admit that this is likely a significant underestimate."[13] These pesticide-related illnesses go beyond temporary sickness. Cancer, loss of toenails and fingernails, recurring rashes, breathing problems—these are just a few of the health issues field workers develop. Pregnant women, who must continue working or lose their jobs, experience miscarriages or give birth to children with physical deformities or developmental problems. Estabrook documents cases of pesticides being sprayed directly on pregnant workers in the field.

In the few minutes Ryan and I spend talking, the crew moves about twenty yards away from us down the rows. I hear Spanish spoken between them occasionally; otherwise the field is silent except for the crunch of lettuce underneath the men's feet. I wonder how many have missed their parents' funerals. I wonder where they live, if they are sick, where their last meal came from. Discarded lettuce heads lie in their wake, left behind due to deformities or disease, looking like oversized green succulents. I see a barely attached head wobbling in the strong wind and bend over to rock it gently under my palm. It feels like it could roll off the bed if I pushed just a bit harder.

5

The Muck

The muck soil is the Everglades Agricultural Area's defining feature, the black gold that drew northern farmers like Ryan's grandfather, Ray, to Florida and keeps their descendants here today.[1] Muck, however, is an exceptional type of soil. Though the name might conjure a soupy bog, muck, or muckland, is so named because of its high percentage of organic matter, 20 to 30 percent by weight. Organic soil components include fresh plant residues, small living soil organisms, decomposing or active organic matter, and stable organic matter called humus. Muck forms under anaerobic conditions, which means little or no oxygen is present. Without oxygen, plant decomposition occurs slowly and organic material accumulates over time, forming soil. Where do such conditions exist? In wetlands, marshes, or other low-drainage areas, such as the Everglades, the draining of which formed the EAA. It takes about five hundred years for just thirty centimeters of muck to form.[2] Though one can find muck in many states, it is not a common soil type, comprising less than 1.6 percent of U.S. soil. Globally, just 1.2 percent of the world is muck.[3]

It's good that the world isn't too mucky, though, because muck suffers from a debilitating problem: it disappears when exposed to oxygen (i.e., when drained). The scientific word for this disappearance is *subsidence*, meaning the soil subsides and becomes shallower. Subsidence occurs in muckland fields because the organic matter decays much faster when exposed to air than when submerged under water. Wind and water erosion and fire also speed up subsidence. Tillage doesn't help either.[4] It's important to note that subsidence occurs in uncultivated wetland areas, too, like during the dry point of an annual flood cycle—anytime muck soil touches air, organic matter decomposes. But subsidence was minimal before humans drained the Everglades, and because human structures like canals have replaced water sheet flow and flood cycles, soil loss is a problem all over Florida. Because far less water reaches the Everglades today than in the past, subsidence is also occurring in Everglades National Park and other preserved tracts.

Wetlands drained for urban development are also paying the price of subsidence: buildings constructed at what was ground level decades ago and anchored in the bedrock now teeter several feet above the ground. Ironically, the buildings at the Everglades Research and Education Center (formerly the Everglades Experiment Station) outside of Belle Glade provide a wonderful visual of subsidence. Stairs now cover the distance from the front door of one building to the ground below, and a latticework cover skirts the gap beneath the structure.[5] Nearby, a nine-foot concrete post measures the subsidence. Someone drove the post into the limestone bedrock in 1924, leaving barely any of the post visible above ground. Today, the post proves that about seventy-five inches of soil, more than six feet, have disappeared since.[6]

Driving across Roth Farms, I see white and gray rocks scattered in the soil. Underneath the Florida muck is a bed of carbonate rock composed of limestone and dolomite known as the Florida platform, the source of the rocks.[7] I ask Ryan if it's common to have so many rocks in the fields. He says no. "The muck is going away," he explains. "It's getting shallow." He shows me a ditch bank that runs along the edge of a field. In this cutaway, I can see the layers of earth: muck, muck sprinkled with white rocks, and white bedrock. The space between the bedrock and the soil surface is about twelve inches. Ryan says this ditch bank illustrates how shallow the muck is becoming—subsidence at work. With fewer feet of soil covering the bedrock, tillage machines now dredge up rocks of all sizes, mixing them with the muck.

When Ray was growing vegetables in the 1940s, the farm's muck was about six feet deep, Ryan says. "We're probably down to about two here now," he says, pointing to a field. On most of the farm, the muck ranges between one and three feet. On a bit of land nearer to Lake Okeechobee, it's closer to six feet. "When you run out of muck, maybe you become a rock quarry, I don't know," Ryan says, only half joking. As much as nine feet of muck is gone in some parts of the EAA. Just over an inch of soil slipped away each year across the region until recent management practices slowed the rate.[8] What's the shallowest muck you've heard of, I ask Ryan later, after I've discovered these numbers. "There are farms that are scraping the muck off one block to put it on another to double the depth of that block," he says. "They are dealing with less than six inches of muck."

Soil loss on farm ground is a national problem, not just a Florida problem. While muck recedes like the tide, other soils blow or wash away. Iowa, one of the most heavily farmed states, had in the nineteenth century fourteen to sixteen inches of topsoil (not total soil, but the upper layer with a high percentage of organic matter and nutrients above the densely packed subsoil). Today Iowa has six to eight inches of topsoil, even less in heavily eroded areas.[9] It's fair to say most of Iowa has half or less of the topsoil it did 160 years ago.[10] Soil loss has occurred far more rapidly in the EAA, with a higher percentage of soil—roughly two-thirds of the original muck—vanishing in just sixty years. At least the muck loss here isn't as bad as in the Sacramento–San Joaquin Delta in California, the country's other major muckland. Three inches of soil vanish in that delta each year—more inches lost annually than in the EAA, but the delta started out with up to sixty feet of muck, while the EAA never had more than twelve.[11] The bottom line: the EAA is shrinking dangerously close to the carbonate bed underneath. Once the muck is gone, there will be only bedrock left.

When soils of any type subside, they lose nutrients and require more fertilizer every year. Farmers once used little or no synthetic fertilizer because of the soil's nutrient-rich organic matter—but heavy fertilizer use is now required on most conventional farms. Ryan gives me an example of how practices have changed as soil subsides: "You never used to side-dress cabbage. Side-dress means putting liquid nitrogen fertilizer into the bin when planting. You used to not ever do that with cabbage. Cabbage is a very low-maintenance crop. You used to just plant the cabbage; you put on dry fertilizer at planting, and then you would grow the crop and take care of it as needed for disease or pests. But fertility was not an issue. Well, they started having issues; the cabbage wouldn't head up properly. It was having nitrogen deficiency. Well, in this nitrogen-based soil we've never had this problem before. Sometimes you have to break free and say, 'It doesn't matter what you used to have, this is the problem of today.'" Farms bear new costs like this, says Ryan, because buyers don't increase their prices if the farmer's input costs are higher. "It doesn't matter to the consumer that you have to [side-dress cabbage] now, and you can't say, 'We have to charge more for cabbage.' No, cabbage is market price. If the market is eight dollars a box yesterday and today we have to put nitrogen fertilizer on, it could still be eight dollars. It could be seven dollars. It could be cheaper and cost us more. You have no control over that whatsoever."

market price

Farmers are doing what they can, Ryan believes, to make the muck last. EAA farmers have employed best management practices (BMPs) to reduce subsidence, such as maintaining a higher water table by pumping less water from fields, incorporating rice and sugarcane into crop rotations (these crops tolerate flooded conditions, which helps slow subsidence), and tilling less often. These BMPs have slowed subsidence from a traditional average of one inch per year to roughly half an inch per year.[12] A few years ago, Roth Farms started collecting radish, bean, and corn culls into a spreader machine and scattering them across the blocks. Eventually the culls break down into the soil, providing more organic matter. "We don't know how much we'll gain by doing that, but we're going to do it," Ryan says. That's probably a good strategy, because if the muck is only twelve to thirty-six inches deep, as on Roth Farms, it will be gone in just twenty to sixty years at the current subsidence rate.

Spreading culls, growing rice, and eliminating turfgrass production definitely help—those are sustainable practices—but such measures only go so far. One or two well-intentioned acts, done in isolation rather than as part of a holistic philosophy of sustainability, barely mitigate the damage caused by other decisions. For example, the cattle manure my father scatters on his wheat fields once a year, while beneficial, isn't enough to make nitrogen fertilizer unnecessary because he doesn't employ a range of sustainable farming processes, such as cover crops and diverse crop rotations. Such is the situation on Roth Farms. Farming conventionally automatically means they take more from the land than they return, no matter how closely they follow best management practices. It's not that they want to take more, or take more because they are greedy or uncaring—they have to in order to survive.

Farmers in the EAA often find themselves at the receiving end of people's outrage over runoff pollution, wetland drainage, and soil loss, which puts them in the strange position of being hated for their practices but loved for the cheap food they produce. While enduring withering public scrutiny, these farmers have watched as urban construction creeps—rushes, in many cases—closer to their farms, bringing with it water-hogging golf courses, houses, shops, and swimming pools, pesticide-laden lawns and golf greens, and pollution-causing traffic. They've seen Florida's beaches erode and its wildlife and sea life populations plummet because of damage to dunes and reefs, all while construction continues on beachfront mansions and resorts. They've heard of city officials granting Celine Dion permission to build a private water park on Jupiter Island just up the road when water supplies in Florida are critically low and their own water use is under the public microscope. The double standards aren't lost on Ryan. Part of the problem, he says, is that most people don't actually know much about agriculture, but they criticize anyway. "It's interesting to me that people have no training, no education in agriculture, and no experience, but they have a ton of opinions about agriculture and how food should be produced," Ryan says. "It's fine; I don't have a problem with it. But most people don't know a farmer."

"Though all professions have their problems, I think farmers are in a tough spot," Ryan adds later. "We have suppliers—fertilizer, fuel, seed, crop protection materials, packaging and labor—that can name the price you must pay for the products they supply, and we have no ability to charge a price for our products that will be guaranteed to pay all of our bills. There are days that we sell our crop for less than the cost of harvesting. And for the privilege, we are criticized around every corner for being Big Ag, lining our pockets with cash while destroying the environment. Farming is a struggle in every sense of the word: we can't control the freeze or the flood, can't name the price we want for our crop, and have to fight to keep doing it."

Ryan sees large operations and industrial practices as the only way to supply consumers with the cheap food they demand. "Everybody seems to want smaller local producers," he says. "But they can't keep up. It's unfortunate. I think it's not the best development for agriculture operations to get bigger, but it is what we're dealing with." There's a note of helplessness in his words. This is the heart of America's agricultural problem: *conventional farmers feel that they have no choice.* It's get big or get out. Many were born into the system and don't know how to farm without inputs—they've lost ten thousand years' worth of agricultural knowledge in a couple of generations. Regaining it doesn't happen overnight. Agribusiness dogma has made them believe that organic, sustainable agriculture is at best a trend, at worst a threat to humankind's survival. They are misinformed about their practices and how to measure efficiency. Many believe, in their heart of hearts, that what they are doing is good and safe.

These farmers don't go industrial, use agrochemicals, or grow GM crops because they are bad people who want to hurt others and the environment. Rather, most are good people trapped in a bad production model, like my father and many of the men and women I interviewed for *Tri-State Neighbor*. Like Ryan, who is beyond a doubt a good man who does not want to harm anyone. I hear it in his voice, and I see it in his life. Like the car seat in the back of his Ford for his daughter, Ella. "She likes the tractors and the big trucks," he tells me. "It's funny that she likes the heavy equipment the most. As she gets older, I'm sure she'll be more interested in the crops, but not today." A picture of her smiling is his cell phone's screen background. "She's amazing," he says. "She's perfect." Ryan is a good father who works hard for his daughter's sake. He's a conventional farmer, but he's not the demon critics might make him out to be.

This is why fixing our broken agricultural system is at once extremely complex and completely doable. It does no good to blame conventional farmers entirely or write them off as evil. It's much better to include them. They should be held responsible for change, but given the tools and support to achieve it. Our agriculture should use empathy and approaches tailored to individual people and environments, because there is no one-size-fits-all agricultural solution. We have tried that approach and it has failed. We also know that changing hearts and minds is possible. That's why I believe we can create a new system—rather, we can *continue* creating a new system, because a number of people have already started blazing a trail forward.

PART TWO
Holistic Regenerative

6
The Holistic Philosophy

In my parents' black diesel-powered pickup truck, I zoom down a narrow gravel road in western South Dakota on my way to Great Plains Buffalo Company, where Phil and Jill Jerde and their family raise grass-fed bison. They do this according to a philosophy called holistic management, which sounds like it could be the name of a corporate seminar. I have only a vague idea of what holistic management is, but I heard it regenerates native grassland and produces practically organic, input-free meat—so I'm on my way to see for myself. The Jerde's herd, over a thousand bison strong, is one of the largest commercial herds in the United States. About 2,500 other U.S. ranches include buffalo, keeping alive a species that almost vanished due to overhunting.[1]

It is early May, fifty degrees and sunny with only a slight breeze, rare stillness in this windy country. I wish it were summer because that's when this part of South Dakota comes alive. A few sights a driver is sure to see from June to August: grassland, cattle, horses, wheat fields, corn fields, hay fields, hay bales, farmsteads, deer (mule deer mostly, especially in the evening and early morning), the non-native state bird called the Chinese ring-necked pheasant, grouse, sparrows, tractors, maybe a pronghorn antelope, pickup trucks, barbed-wire fences, brown hawks, rocks piled at the edges of fields, roadkill, wild sunflowers, white country churches, and people riding horses.

Instead I see empty or just-sprouted fields and pastures with no cattle.

Spring comes late to South Dakota; May is often cold and has been known to bring snow. The pastures show a hint of green beneath last year's dried grass, but the new growth is minimal: this area—and much of the Great Plains from North Dakota to Texas—is suffering from drought. The U.S. Drought Monitor map from that week shows the northwest corner of South Dakota colored in bright red that signifies the "Extreme" category; not as severe as the burgundy "Exceptional," but dry nonetheless.

After 6 miles or so, I turn on a narrower, even less traveled road. By now I'm roughly 85 miles from the nearest fast food restaurant, 40 miles from a high school, 20 from a gas pump. This is remote, even for someone who grew up 10 miles from the "McFarthest Spot," the longest distance from a McDonald's in the continental U.S., which is 145 miles by car, 107 miles as the crow flies.[2] Phil said there would be buffalo skulls and a fire hydrant at the turn-off to his ranch. This is how ranchers give directions. They tell you things like, "Go . . . well, I don't know, about 5 miles down the road, and then you see the big cottonwood tree on the north side and go another mile or so after that and turn by the pile of rocks. You can't miss it." Right.

Found 'em: skulls and hydrant. I make a quick left onto a still narrower road and reach a large log cabin on a hill. A pickup emblazoned with the words "Great Plains Buffalo" tells me I'm at the right place. After visiting with Jill and all but one of the ten Jerde children—who range in age from newborn to twenty-one—Phil says it's time to see the buffalo. Phil, forty-five, looks like a typical high plains rancher. He wears a long-sleeved white shirt with olive-colored checkers, dark blue Twenty X brand denim jeans, and what I'm sure is a hand-tooled leather belt, a common cowboy accessory. A short-trimmed beard covers his chin, cheeks, and upper lip. One unusual accessory, though: a baby carrier, the cloth kind that hooks in the back and allows the wearer to carry the baby in front, against the chest. Inside the carrier, baby Quilla sleeps, her blonde peach fuzz hair just visible. Phil loves taking his children along to do ranch work, especially the babies because he says they sleep well in the carrier (how I don't know because ranch work is rarely quiet). "At this age they spend more time with me than Jill," he says proudly later.

Phil, baby Quilla, and I get into a ramshackle blue Ford designated as "the fencing pickup." Jesse, age four, and Jack, age six, scramble into the truck bed, which is littered with fencing material: a wire stretcher, barbed wire, a steel post driver, fencing staples, wire cutters. I am amazed because my family also has a '70s-era blue Ford fencing pickup almost exactly like this one. We call it Old Blue.[3] Fencing pickups are generally old trucks no longer reliable enough for highway use, and they serve the dual purpose of storing fencing supplies and transporting people around the ranch. Dents, torn seats, low-functioning brakes, touchy gas pedals, and tricky ignitions are common. My family's fencing pickup has left me stranded more than once.

We head for the pasture, off to do a trial run of a new portable water delivery system Phil has devised for providing water to the buffalo herd as it moves from one pasture to another. He's chosen a long, rectangular steel trough to act as a water tank, and to that he welded an even larger flap of durable rubber. To the rubber flap he attached a chain so he can drag the tank to new pastures. Whenever Phil needs something for the ranch, he first considers whether he can build it. If no piece of equipment exists to meet the need, as with this water system, then he often invents what he's looking for, welding it to life in the Quonset machine shed.

After building the water tank, Phil had to figure out how to pipe water to it. He bought thousands of feet of above-ground water pipeline made of plastic so strong "a bulldozer could drive over it," he says. He cut two sections of pipeline for today's test run and fastened on special connectors so the pieces can be spliced together as needed. Extra pipeline wraps around a six-foot reel mounted on a trailer—it looks like a huge spool of oversized thread. When the buffalo herd moves to a new pasture, Phil will connect the appropriate length of pipeline to the closest water well (there are several dotting the ranch) and then stretch the line to wherever the portable tank awaits.

Most conventional ranchers rely on one or more permanent water tanks scattered throughout sprawling pastures, as well as natural water sources. This forces cattle to constantly return to the same place for water, which limits their desire to wander too far away, especially during the hot high plains summers. There are several reasons Phil wants a system that follows the buffalo instead of remaining stationary. One reason is the drought. It's been three weeks since the ranch received snow or rain, but the overall shortage spans several years. The last summer and two winters have been dry—so dry that the grass growth on Phil's pastures was not abundant enough to support the bison through the winter, so he had to purchase supplemental hay. So dry that the region suffered prairie fires and failed crops. So dry that natural water sources such as creeks, ponds, and dams shrank until some were little more than shallow pools. Unless the summer brings rain, these sources will disappear and "we'll have to do some serious destocking by fall," he says, meaning he'll be forced to sell part of the herd.

The main reason Phil needs a portable water system, however, is not the drought. Droughts occur in cycles on the Great Plains—years of plenty followed by years of want is typical, although climate scientists have proven that recent droughts are more severe and long-lasting due to climate change.[4] Phil needs a new system because of the unconventional way he manages his pasture and livestock, a way that insulates the land from the worst effects of drought that conventional grazing can't. His way encourages grass growth and soil health, discourages overgrazing, and treats the ranch and its soil, animals, and people as one ecosystem—and his way sets him apart from the majority of ranchers.

But before one can fully understand Phil's method, it's important to understand what it is not—and it is decidedly *not* conventional grazing.

Conventional grazing, practiced by the majority of ranchers in the United States, is generally this: ranchers turn cattle loose in large pastures and allow them to graze selectively using "free will," choosing the plants they like best and leaving everything else untouched. Conventional grazing became standard practice on the Great Plains in the 1870s, when barbed wire appeared and cattle barons carved the Great Plains into pastures. Cowboys no longer needed to move cattle to fresh ground and keep them from straying; instead, ranchers could simply put cattle in wire-enclosed pastures, shut the gate, and let them be, a philosophy that hasn't changed much since. Conventional ranchers rotate their cattle to new ground after a pasture becomes "grazed up," as they often put it. This might be once every few months, depending on the size of the pasture and herd.

Not only does conventional grazing require little work, but cattle also perform well under this method. Because they consume the most nutritious grasses and leave the rest, they fatten quickly and rebreed easily. Operating under conventional grazing theory, ranchers opted for large-framed cattle breeds hard-wired for big meat production, breeds with high nutritional requirements best satisfied with selective grazing at low stocking rates (i.e., fewer livestock per acre), as well as with supplemental feeds and chemical parasite control. Conventional grazing can also include spraying pastures for weeds and applying fertilizers—quick chemical fixes for problems caused by this management strategy.

Instead of conventional grazing, Great Plains Buffalo practices holistic management, a theory of land management first espoused by biologist and environmentalist Allan Savory as he sought to understand and reverse the growing desertification of the world's grasslands, particularly in Africa. Savory calls his method holistic because it accounts for the needs of the whole grassland ecosystem—soil, plants, insects, grazers, wildlife, people. Phil firmly believes in Savory's work, and he mentions holistic management right away when I ask what makes Great Plains Buffalo different from other ranches. Savory first presented holistic management and the Savory grazing method in the early 1980s through journal articles and, later, more comprehensively in his book *Holistic Management: A New Framework for Decision Making*. Holistic management is, at its heart, a decision-making framework, and Savory presents several key insights to guide land decisions. First, he says, land managers must realize that "no whole, be it a family, a business, a community, or a nation, can be managed without looking inward to the lesser wholes that combine to form it, and *outward to the greater wholes of which it is a member* . . . in studying our ecosystem and the many creatures inhabiting it we cannot meaningfully isolate anything, let alone control the variables."[5] A good land manager realizes that every decision has visible and invisible consequences in the environment, and no one action or inaction can fix or single-handedly create a problem. This sounds a lot like Joel Salatin, a farmer featured in Michael Pollan's *The Omnivore's Dilemma*, when he's talking about his biological farm with carefully layered enterprises: "In an ecological system like this everything's connected to everything else, so you can't change one thing without changing ten other things."[6] Good farmers and ranchers recognize and work with this complexity.

Savory also calls for a new way to classify environments: on a continuum from nonbrittle to brittle according to the evenness of precipitation and humidity distribution throughout the year and how fast vegetation decays. Seeing land on this continuum explains why environments react differently to the same forces, such as grazing. A nonbrittle environment, or a humid place like the tropics with consistent rainfall and fast decay, responds well to rest or long periods of no grazing because the environment naturally induces the breakdown of dead plants. It doesn't need grazers for that job. But a brittle environment, or a place with little humidity, and with erratic rainfall and slow decay, such as the grassland plains of Africa and America, turns into a desert when rested.

Why? Conventional wisdom suggests that arid or semiarid land receiving inconsistent rainfall should be rested, not grazed. Savory says that in these brittle places large grazers, not humidity, are the engine for recycling nutrients, breaking down vegetation, and keeping microorganisms alive during dry periods. Take those herbivores away, and the grass receives no fertilizer, insects and microorganisms have no food, and no plants decay, resulting in stands of dead grass that choke out new plant growth.

Desertification is the end result. As evidence, Savory points to Africa, where millions of animals once roamed the plains—many times more than the number of livestock animals now present—and they enjoyed a plentiful environment. Yet when comparatively small numbers of livestock replaced wildlife, the grasslands deteriorated. Experts, scientists, and government land managers insisted that the desertification was a result of overgrazing and excessive trampling caused by too many livestock on pastures and too much wildlife concentrated within game preserves. Thinking the same thing, Savory once called for the culling of elephants and buffalo and ridding the African grasslands of livestock altogether. But the research station plots he saw that were "properly managed" with few livestock still became little deserts.

Years of observing wildlife and livestock finally led Savory to conclude that time, not animal numbers, was the key to understanding grassland health. Overgrazing is not connected to the number of animals in the environment, but instead to the *amount of time the environment is exposed to the animals*. If livestock or wildlife keep returning to the same area without giving the plants appropriate time to rest—and "appropriate time" can be short or long depending on where the land falls on the nonbrittle to brittle continuum—then the plants weaken, die, and are replaced by bare ground or unpalatable plants: desertification in action. In a brittle environment, it's all about high intensity, short duration grazing followed by rest. That's why grasslands supported high animal numbers in the past; herding animals, bunching closely as they did in the presence of predators, would produce dung and urine in high concentrations on their grazing area. Animals do not like to feed on ground they've fouled, so they keep moving to avoid it, and also to evade predators. A day or two of hard use or "massive disturbance," as Savory words it, followed by a long period of recovery: that's what ideal grazing looks like in a brittle environment.

At its most basic level, holistic management is similar to the theory of rotational grazing, which means moving livestock frequently to new pastures, grazing each area intensively for a short period of time, and then allowing it to rest. But Savory claims that rotation alone isn't enough, and the insights mentioned previously are only the beginning of holistic management. He covers the water cycle, mineral cycle, energy flow, community dynamics, technology, money and labor, fire, rest, and animal health. An understanding of all of these, he says, is necessary to manage grassland. Savory also remarks on financial planning, sustainability in resource use, society and culture, the purpose of research, and effective policymaking. My favorite is the chapter titled "Cause and Effect: Stop the Blows to Your Head before You Take the Aspirin" from the second edition.[7] It's a book about grassland management, livestock, and desertification, but it's also about living holistically no matter one's occupation.[8]

Savory's theory united a variety of people—scientists, grazing experts, ranchers, government bureaucrats, and environmentalists—on the issue of stopping desertification. Unfortunately, it was usually for the purpose of denouncing his research. Savory's grazing theory is highly controversial to some people, who write that he is eccentric and perhaps crazy, that his ideas actually create more deserts and destroy what is left of the world's grasslands. Others say outside trials of holistic management prove it's no better than conventional grazing, a claim Savory disputes by pointing out that the trials only test rotational grazing, not his entire holistic management approach.

Many ranchers take issue with Savory because his insights challenge the way they've raised livestock for well over a century. They don't like the idea of throwing out generational wisdom and implementing a new approach. Adopting Savory's theory means a paradigm shift in how they operate and major structural changes to their ranches. Most ranchers reject the argument that conventional grazing causes desertification in the first place. While there are some bad apples that overgraze their ranges, ranchers tend to see overgrazing as an abomination. I've heard my father say as he drives by pastures nibbled to the dirt, "Those poor cows have nothing to eat," a comment not only on the sorry state of the grass but also on the rancher's intelligence, which is assumed to be quite low if he or she has let the pasture get that short. As long as they don't overgraze, ranchers tend to think, they're doing things right.

Despite their best intentions, however, many conventional ranchers see undesirable plant species like Canadian thistle and leafy spurge replacing their grasses, prairie dogs overtaking their ranges, and pastures that come back less lush and thick each season. Their land collapses during drought years, forcing them to sell livestock. Some respond by spraying weeds and thistles, poisoning prairie dogs, and fertilizing with synthetic products. Others plow up native grass and try other varieties that supposedly work better. In their minds, the land is fine—it's just outside forces like weeds and drought that need to be overcome. This is why it's tough to convince ranchers that conventional grazing is harmful: they simply can't see the connection between desertification and what they consider to be best practices, which to them means giving livestock huge amounts of space to scatter.

Meanwhile, environmentalists often see livestock as the source of grassland degradation, air pollution, and climate change (and cattle confined in feedlots *do* cause methane pollution, which contributes to climate change). Savory's call to increase livestock numbers—"Only livestock can save us," he has said—sounds like the worst possible solution.[9] Many environmentalists want grasslands to recover on their own, without human or livestock intervention. This hands-off strategy has been implemented on many government-managed ranges across the country, mostly a result of pressure from environmental groups.

Yet we cannot just leave the grassland alone because today's grassland is nothing like it was in the past, when many species of plants and animals fueled a vibrant prairie environment. Humans removed the largest herbivore, the bison, from that environment. An estimated thirty million to sixty million buffalo roamed North America when Europeans first set foot on the continent.[10] Their herds sometimes numbered in the hundreds of thousands—anywhere they stopped would run out of grass and water quickly, so they had to keep moving. And after grazing an area along their seasonal routes, they might not return until the next year. Unbeknownst to them, they practiced holistic management. By 1884 just 325 wild buffalo had survived the U.S. Army–sanctioned slaughter campaign intended to make way for cattle and weaken the Plains Indians by destroying their main food source.[11] The presence of white settlers also reduced other herbivore populations, such as deer, elk, and pronghorn antelope. Removing these herbivores had a ripple effect, since most flora and fauna evolved in response to impact from large herbivores.[12] Bison, however, were the "keystone herbivores" in this ecosystem—without them, the chain of life started to collapse.

Because our forefathers reduced (and we continue to reduce) the presence of wild animals by plowing up the prairie to plant corn, expanding our cities, and polluting the water, *we* are now responsible for the health of the grassland. Whether we like it or not, we are the keystone species, and we are tasked with replicating the bison's effect. Perhaps on government land we can reintroduce buffalo, following a model like the one at Custer State Park in South Dakota or Yellowstone National Park in Wyoming. But that's not an option on privately owned ranch land, which includes the vast majority of our nation's grassland.

If livestock left the land, then most ranges would never experience the effect of herbivores grazing, excreting, and walking, at least not abundantly enough to fuel natural mineral and energy cycles. Leaving the grassland alone, virtually devoid of herbivores, is unwise for the plants, insects, and small animals that depend on symbiotic relationships with them. Brittle grassland environments become deserts without herbivores—some are already halfway there. Holistic management offers a way for livestock to be part of the grassland ecosystem. They can fill the gap in the symbiotic relationship between herbivores, soil, grass, insects, and other animals. Phil shows me a picture that illustrates such a relationship. The photo is of a buffalo hoof print. "That's some bare ground that a critter impacted while we were in there," Phil says.[13] "You can see there are even some seeds lying there on top of the soil. That hoof print is seed to soil compaction. It made a depression so that when we do get a little bit of moisture it's going to tend to funnel it right along that edge, and that's where your new plants are going to grow." The hoof print reveals the complex impact herbivores have on the grassland: they "till" the soil, plant the seeds, help water those seeds, and fertilize the ground.

Surveying the picture, I'm astounded at the intricate and mostly unseen relationship between herbivores and the prairie, a relationship I had no idea existed. No one I knew growing up talked about stuff like this. After several interviews with Phil, I start looking more closely at our ranch's pastures for signs of desertification. They aren't hard to spot: areas of bare ground, invasive plant species, woody plants, and prairie dog towns that cover hundreds of acres. The worst part is how thin and short our pastures are at the height of summer compared to Phil's. When I try talking to my father about how better rotations might help, he's not receptive. I can't blame him. Before my *Tri-State Neighbor* days, if someone had told me our ranch contributed to desertification, I would have scoffed. I used to think I knew everything about cattle and ranching, used to think protecting and regenerating grassland was the same as making sure cattle didn't crop it too short. It turns out I'm just starting to understand how cattle actually affect the land.

7
The Grass

Jesse and Jack, still riding in the back of the pickup, yell at their father to "Stop Papa! Stop!" as we drive past a pile of bleached white buffalo bones. Phil says out the window, "We'll come back here, boys." He turns to me and grins. "They are really into skulls."

Only farm kids can be really into skulls, or grass for that matter. These children can identify different grass species and explain the role of dung beetles in the nutrient cycle. Most ranchers do not have the knowledge that the Jerde children already possess. Phil certainly didn't in the past. He first learned about holistic management and Allan Savory over a decade ago at an informational meeting for ranchers hosted by the Cooperative Extension Service. His daughter Emily, then eleven, and son Payton, nine, were with him (even then Phil brought his children everywhere). It was a meeting that changed the course of the ranch's future.

The message he heard—that overgrazing isn't about animal numbers, but the amount of time the grassland is exposed to animals—was a major revelation. Anxious to learn more, he bought Savory's book. Phil tells me, "They said [at the meeting], 'You won't be able to read this book because it's too boring of a read.'" Note: Savory's book is more than six hundred pages long in small type with few pictures. But Phil is an avid reader. "It didn't take very long and I read it, and I read it again," he said. "It was like a light turned on."

Over the last decade, the Jerdes have divided the ranch into one hundred pastures of roughly 130 acres each. Phil forces the buffalo—1,000 to 1,200 of them, depending on the time of year—to graze these pastures to the point where ranchers would accuse them of being overgrazed. Savory's "massive disturbance" happens, then Phil moves the buffalo to the next pasture and doesn't allow them back again for a year or so. The Jerdes play what Savory identified as the role of predator because they push the bison to move every few days. While watching buffalo herds in Africa, Savory noted that animals walked gently and slowly when spread out, placing their hooves beside and not on top of coarse, inedible plants. They also placed their full weight on the soil, compacting it. When the same animals moved quickly and bunched together in new grazing areas or when they fled predators, the impact on the environment changed:

> I noted that while bunched as a herd animals stepped recklessly and even very coarse plants, containing much old material that would not be grazed or trampled normally, were trampled down. That provided cover for the soil surface. In addition, the hooves of bunching and milling animals left the soil chipped and broken. In effect, the animals did what any gardener would do to get seeds to grow: first loosen the sealed soil surface, then bury the seed slightly, compact the soil around the seed, then cover the surface with a mulch.[1]

The impact of bunched animals is vital to the land's health, not harmful as people perceive it to be. Because livestock have few predators today, ranchers need to re-create the herd behavior. And if there's anything buffalo really enjoy doing, it's running as a herd. A few months later, Phil shows me video clips of the herd galloping into fresh pastures, snorting and kicking their hind legs in the air with glee. They're nimble and quick, dashing up hills and cutbanks on their skinny legs. In one video they charge through a stream, water splashing up around them. When I study them on that May morning as Phil and I drag the pipeline, I notice how they run almost everywhere: to the water tank, away from it, toward the pickup, back out to pasture. The young ones are like big hairy puppies, playfully jumping around. I can see why the buffalo shaped the prairie—they're in almost constant motion.

The result is even distribution of animal impact, a major departure from conventional grazing, where animals return to the same areas frequently while leaving others barely touched. As even conventional grazers admit, livestock will eat choice plants first and leave behind woody, unpalatable plants. Because of this selectiveness, livestock left alone in massive pastures eventually overgraze and kill the nutritious plants, allowing woody plants to take over. Conventional grazing also results in uneven mineral distribution, in the form of manure and urine, across a pasture. "Typically what happens in big pastures is the animals go back to the water source where they drink, lie down nearby, stand up, pee and poop, and then they go out and graze again and repeat the process," Phil says. "So what ends up happening is you move all the minerals from where they're eating back around the water. Under planned grazing, you end up with even distribution of minerals back on the land. When we drive across here, I want to see manure patties everywhere, even, which I think we're seeing."

He motions toward the pee and poop evidence outside the pickup window as we traverse the pasture and, yes, there are manure patties plopped consistently over the land. We work as we talk: we hitch the portable water tank to the pickup, pull it to the pasture where the buffalo are, go back and hitch up the waterline, and pull one end to the tank. I open any gates we encounter. When he stops the stick-shift pickup and it rolls just a little, Phil sometimes says "Whoa," as if to a horse. "So, short duration, high intensity, followed by a long recovery," he continues.

High intensity is right. Phil tells me he stocks his land at about one buffalo per fifteen acres of land.[2] Under conventional grazing, a rancher might be able to stock one beef cow (which eats less than a buffalo cow) for every twenty-five to thirty acres of land in this part of South Dakota, he says. Phil's land supports twice the livestock—yet his pastures contain far more forage than conventionally grazed land. The buffalo graze year-round, even in the dead of winter, something a conventional ranch couldn't support. How can this be? Overgrazing is less about what the grass looks like above ground and more about what the roots look like underground, Phil explains. "Savory writes how you really can't overgraze a plant in three days. Overgrazing is a matter of time, not a matter of animal numbers. When a plant is grazed off, and it's starting to shoot up again like you see these plants out here"—he points to the shoots of grass as we drive by—"they're doing that on root reserves. Until they get big enough to where they have their solar panels out, they're all going to have to borrow energy from the ground." He pauses. "There's a root die-off going on, is what's happening." He means in the region, on the grassland as a whole. I can see from his expression that a root die-off is troubling. "So if I continue to graze that plant off before it can get up and start harvesting solar energy again, I deplete that root mass until eventually I kill that plant."

A conventional pasture might look verdant, but the root mass underneath is likely depleted, because each time the grass launches new growth, livestock eat it. This happens all summer long, year after year. The plant never has time to fully launch its "solar panels" and restore what was taken from the root mass. It keeps growing, but on borrowed energy. Eventually the plant stops sending shoots to the surface because its roots are exhausted. That's one reason pastures in the Great Plains are slowly becoming thinner, less diverse, more populated with weedy and woody plants, and more at risk of desertification.

The evidence for the desertification of America's grasslands is startling. But first, what is meant by desertification? One group of scientists define it as "the loss of the ability of a landscape to provide ecosystem services that are important to sustain life," like biological or economic functions.[3] Bare ground exceeds vegetation cover and nutrients stop cycling. Desertification is not simply an expansion of a desert, though; it can occur anywhere, even far from an existing desert. Other scientists define desertification more widely or more narrowly, but most agree that desertification is caused by a number of factors acting simultaneously, primarily human activity (i.e., poor management such as overgrazing and over-farming), higher than normal temperatures, and lower than normal rainfall. The International Fund for Agricultural Development says about 29.7 million acres of land are lost to desertification around the world each year, which threatens the livelihoods of about one billion people in more than one hundred countries. The group also reports that 25 percent of the world's land is desertified.[4]

Grasslands and other dryland environments are especially prone to desertification as climate change accelerates.[5] Here in the United States, desertification is already happening. A joint report from the Rocky Mountain Climate Organization and the Natural Resources Defense Council describes larger wildfires, shrinking rivers and snowpack, and the loss of forests, glaciers, and wildlife, all results of climate change.[6] They note that ranchers are suffering because of desertification, with drought-related herd culling, lost profits, lack of feed, and stressed water sources.[7] The report also reveals another disturbing fact: the American West is growing hotter and drier at a faster rate than anywhere else in the nation and in some cases the world. From the report: "For the last five years (2003 through 2007), the global climate has averaged 1.0 degree Fahrenheit warmer than its 20th century average. For this report, RMCO found that during the 2003 through 2007 period, the eleven western states averaged 1.7 degrees Fahrenheit warmer than the region's 20th century average. That is 0.7 degrees, or 70 percent, more warming than for the world as a whole."[8] In other words, the United States stands to experience some of the worst desertification caused by climate change in the world. A more recent study, the 2017 Climate Science Special Report, confirms the RMCO's work, showing that temperatures in the northern Great Plains are, on average, 1.69 degrees higher now than in the first half of this century. The report also describes fires, heat waves, melting snowpack, severe storms, floods, and droughts caused or worsened by climate change.[9]

What can ranchers do? In Phil's eyes, reversing desertification primarily means preventing bare ground. "Bare ground is the worst thing because water doesn't penetrate bare ground very well and water likes to evaporate off of bare ground pretty rapidly, too," he says. A few months later, he illustrates the harmful effects of bare ground by showing me pictures that he took after "a big rain event." More than four inches of rain fell during that summer storm. A picture of the neighbor's side of the fence showed a clear path of dirt and matted grass where the rainwater had formed a miniature river. The neighbor's land absorbed very little rain; it ran off instead because the conventionally managed pasture had bare ground and thin grass. Thistles, dried grass, and dirt clung to the fence where the water had flowed under. On the Jerde side, though, the mark of the miniature river disappeared. Phil's land absorbed the runoff completely. Phil showed me another picture from a month after the big rain: his neighbor's pasture still bore the scars of erosion, while his pasture was especially lush and green where the water had flowed in. Phil turns to me and asks, "Is it a problem of not getting rain, or not using the rain that we get?"

To Phil, people on the land like him and his family are responsible for keeping it healthy so it can, among many other ecological tasks, absorb rain. The stakes are high these days. Because desertification threatens food production, it also threatens national security. "A society that can't feed itself soon goes to war to get food," Phil says. "Throughout history this kind of stuff goes on. The ones on the land are the ones responsible for whether the land continues to be productive or goes backwards." At first I think Phil is being overly dramatic. Flawed as our agriculture is, the U.S. will surely never need to wage war over food, right? Not so fast. Later, I come across Dan Barber's discussion of the "law of return" in his book *The Third Plate*. The law of return, put into words by the father of organic farming, Sir Albert Howard, is based on the fact that soil regenerates itself through the circle of life: death, decay, regrowth. This circle creates what Barber calls "a long-term bank account that provides for the future needs of plants."[10] Farming extracts soil fertility, which is not inherently bad, but it means we have to put back as much or more fertility as we take—the law of return. If we don't put anything back in the bank, the soil can't regenerate, which is what happens in conventional agriculture today.

It turns out that past civilizations also neglected to pay the bank. Barber reveals that empires such as ancient Rome, Greece, and medieval Europe "built their success on the same system of careless banking. They grew food and transported it long distances to feed a growing population. They cashed in on the fertility without paying back the bank. This worked for a while, but ultimately the soils stopped producing." We know what happened next: those civilizations eventually collapsed. These historical moments should be warnings to us about the danger of forgetting to pay the soil bank. But instead, we throw away the past-due notices and keep on going.

Holistic management also reverses desertification by increasing plant diversity. Phil says the many types of grass in his pastures thwart the logic of conventional management wisdom, which says this section of the Great Plains can only grow one type of grass: cool-season grass. "We've been told that this is cool-grass country, cool-season grasses," Phil says, still with baby Quilla strapped to his chest.[11] "Well, what we're finding when we manage with this short duration, higher-intensity grazing is that our warm-season plants make a return." The prairie once supported a diverse mix of cool- and warm-season grasses. In *The Prairie World*, author and ecologist David Costello counted in eastern Colorado's prairie 143 species of forbs, 22 species of grasses, 10 varieties of shrubs, and 4 kinds of trees—and this was in 1969, when conventional grazing had been the norm for almost one hundred years already.[12] Eastern grasslands, like those in Illinois and Iowa, once had even more variety than the short-grass western prairies. Having a wide variety of species meant something was always growing, whether in cold May or hot August. By July, cool-season grasses were gone or became too tough and dry for livestock to eat. Warm-season grasses took off at this point, providing forage for the hot months.

Conventional grazing, however, caused a gradual shift toward more cool-season grasses because livestock return again and again to the same warm-season grasses as the summer wears on.[13] With no time for regrowth, the warm-season grasses rely on their root masses for nutrients, a losing battle because every time the plant grows a little, the livestock come back. After years of root-mass depletion, they eventually stop sending up shoots, allowing cool-season grasses to take over permanently. Seeing that their pastures are thin by July, many ranchers accidentally exacerbate the problem by introducing non-native cool-season grasses, such as crested wheatgrass, to boost forage. Cool-season grasses seem like the ideal fix: the seed is cheap, they are easy to establish, and they green up early in the spring so ranchers can turn livestock out to the pasture earlier. If rains come in the fall, then cool-season grasses will green up a second time. But when the summer heat hits, as it always does, these grasses retreat, and on most conventional pastures there are few native warm-season grasses waiting to replace them. Weeds and woody plants such as yucca and sagebrush move in, bare ground emerges, and wildlife move elsewhere.

Phil has coaxed warm-season grasses back to his ranch, another reason the pastures stay so lush through the hot South Dakota summers. "Warm-season grasses, one way to look at 'em is, they'll use half the moisture and produce twice the forage," he says. Cool-season grasses tend to be short and lose their nutrition early in the season, while warm-season grasses tend to be tall and highly palatable with lasting nutritional value. They provide better erosion control, more cover and forage for wildlife, and higher fire tolerance, too. During times of drought, having warm-season grass is especially beneficial because they have deeper root systems to access water, as Phil witnessed during the bad drought years of the early and mid-2000s.[14]

The deep root systems also mean that warm-season grasses are a more complex (read: desirable) nutrition source for the buffalo and, in turn, for the consumer. "A lot of these [warm-season] plants, they have roots that go down fifteen, twenty feet in the ground," Phil says. "They are accessing different minerals and bringing them up and making them available to the animals, whereas when we have monocultures of crested wheatgrass or monocultures of whatever, we only end up accessing minerals at a certain depth of the soil at a certain time. We want diversity; we want many different warm-season grasses, cool-season grasses, warm-season forage, cool-season forage."

Phil isn't targeting any particular species of grass. As long as his pastures contain a diverse mix, he says, he likes any grass that will grow and provide good nutrition for the buffalo, native and non-native alike. "Most of this is native," he says as we drive through a pasture. "And now we're driving though the smooth brome—well, that's an introduced grass. A bunch of the stuff we have is crested wheatgrass, which is another introduced grass. Native, non-native, like Savory says, the non-native species are just ones that didn't get here soon enough to get their green card." Phil chuckles. "We arbitrarily assign that title to them. Most of the ones people consider invasive, non-native, are really just a function of our management. If we are going to manage for cool-season grasses, that's what we'll get."

Many ranchers think the grassland can't be changed to grow different things. What grows in the pastures is just what grows in the pastures, period. Most don't see the connection between conventional grazing decisions and the near monoculture of cool-season grasses, or don't believe it when they hear about it. One reason for this disbelief is that most ranchers were not alive to see the grassland as it used to be: diverse, with warm- and cool-season grasses swaying waist high and lasting all summer. Cool-season prairie is all they know. Phil, on the other hand, sees the grassland differently—as part of an ecological whole in which he can't manipulate one thing, like introducing cool-season grasses, without impacting everything else. In the natural world, one action or input does not yield one result or output. He can't simplify nature's diversity, something industrial agriculture seeks to accomplish with monocultures, a militaristic focus on yields, and human-made substitutions for natural processes. He views himself as a member of the ecological whole, meaning he is responsible for its well-being. This whole extends beyond the ranch—it's the entire world.

So is holistic management working? On Phil's ranch, holistic management has helped heal the land from decades of prior conventional management, bringing it closer to what it was in presettlement days. "It's pretty amazing over time what can be done, in pretty fast order, too," he says. The ranch isn't 100 percent restored—that will take many more years—but the Jerdes are on the right track. Phil has seen dozens of native species return and bare ground give way to vegetation. On land he purchased in neighboring Harding County, for example, Phil noticed a rectangle of bare ground in the midst of the pasture. "You drove out there and saw it; you saw the straight lines. I think that that was the original homesteader's piece of ground he dug up. A hundred years ago, we did something, through our management, to the soil. That guy went broke and moved on, but under continuous grazing, season-long grazing, that land was never quite healed. Now you go out there and you have a hard time finding those lines that were clearly visible eight years ago."

We drive into a low spot near a creek through a small patch of "buck brush," as ranchers call a certain perennial, shrub-like plant with woody stalks and small, oval leaves. Buck brush grows in thick stands about two feet high, and it can take over low-lying areas. This stand is small, less than ten feet across. "We had lots of buck brush out on that Harding County land, which cattle don't like. It's obvious," Phil says. "But yet this is where the most moisture in the pasture is. It should be our most productive piece of land. It's being wasted really. What we would see over there, is when we were in a pasture for a short period of time, good manure and urine distribution, pretty soon the species started changing. Warm-season, waist-high, chest-high; big bluestem, switchgrass, instead of buck brush. It's pretty exciting stuff."

Better management has also encouraged more animal biodiversity. "We have way more sharp-tailed grouse and partridge," Phil says. "Deer and antelope not so much of an increase; we had a lot of them anyway. The birds are the big one." Invisible creatures are just as important as visible ones. "When we go move buffalo, I thought we were moving a lot of buffalo. But it didn't take very long and I figured out we were moving way more birds—there were thousands of cow birds following the herd," Phil says. "Then pretty soon I figured out, well, the birds are following because there is all kind of insect life, dung beetles mainly. I figured out we were moving way more dung beetles than we were moving birds when we moved the big herd."

Why do the dung beetles matter? In Phil's eyes, grassland operations such as nutrient, water, and carbon cycles perform best when animals like dung beetles and birds do the work. He shows me a picture of a manure patty that he opened up with a shovel—yes, Phil documents the contents of buffalo poop—and I can see the small, brown dung beetles. "There are three different species in this one pile. These guys are taking that manure and taking it down and burying it in the ground. If you put Ivomec on the cows, they quit working for you.[15] It kills them. These are the most common ones, the little red ones. You can open a pie up and find one hundred to a thousand in each one."

The dung beetles move the nutrients from the manure back to the soil, so more grass can grow and continue the cycle. The birds help the buffalo by eating insects that can carry disease and be a nuisance. The birds also leave droppings on the ground, another variety of fertilizer for the grass, and they eat and poop out plant seeds, which means they assist in creating diversity on the grassland. There are probably many more unseen benefits from this symbiosis that add up to healthy grassland. Fascinating enough—and then Phil discovered he was moving creatures even smaller than dung beetles. "It turns out there's these little mites that ride on the backs of the dung beetles, and when they get to a new patty then they jump off—and they eat fly larvae, that is what they're going after, fly eggs—and so when it's time to leave they jump back on the back of the dung beetles and head to a new pie. It's a symbiotic relationship because the fly larvae are competing with the dung beetles for that manure space. It's a win-win for them."

Diversity from the micro level to the macro level: that's what agriculture is all about for Phil. In the agricultural world, diversification has two meanings, the first and simplest being that the operation includes either a combination of crops and livestock or many different species of crops or many species of livestock. The second, more complex meaning includes the environment. Diversified farming systems (DFS) are, collectively, an agricultural model that "share[s] much in common with sustainable, multifunctional, organic and local farming systems, but are unique because they emphasize incorporating functional biodiversity at multiple temporal and spatial scales to maintain ecosystem services critical to agricultural production. These ecosystem services include but are not limited to pollination services, water quality and availability, and soil conservation."[16] Diversification in this sense means welcoming nature into the farm model as Phil does, using its free services to accomplish tasks like fighting disease or controlling weeds. An adjacent marshland helps with flood control, or a section of forest provides natural habitat for wildlife that fertilize the farm's soil, or birds and mites control insects on buffalo.

Pest control through diversified farming contrasts sharply with conventional pest advice. The University of Nebraska–Lincoln's Beef Division, for example, offers the following for controlling flies: dust bags filled with insecticide powder that the cattle rub against to dispense powder onto their backs; back-rubbers (oilers) filled with oily insecticide that operate in a similar fashion as dust bags; animal sprays that the rancher or a hired contractor mists on the livestock, oral larvicides (feed additives) that the cattle consume in the form of loose mineral, lick blocks or lick tubs; pour-on insecticides like Ivomec applied to the cattle's backs; and insecticide-coated ear tags. UNL Beef also recommends a device called the Vet Gun that shoots capsules of insecticide, which burst open when they hit the cow. Every product is connected to the agribusiness supply chain. There is no mention of natural fly control methods.[17]

Recognizing and encouraging diversity and its many symbiotic relationships saves Phil time—and money. The ranch's input costs stay low, a financial boon because his bottom line improves and his business is insulated from input price jumps or buffalo market crashes. He doesn't pay for feed, vaccines, pest control products, antibiotics, or fertilizer. He doesn't spend money or time raising corn, hay, or other crops for feed. As a result, Phil is more independent. He relies on himself and nature, not agribusiness.

I see the evidence that holistic management works. But isn't Phil's system a kind of buffalo monoculture? Shouldn't ranchers focus on many species instead of just one to better replicate the original grassland environment? In some parts of the world, yes. In others, not necessarily.

In a 2009 study, researchers analyzed three types of grassland in South Africa and three types in Kansas: one with multiple herbivores at work, one with a single species, and one with no herbivores. The land with no herbivores experienced a high loss of plant diversity, with woody, unpalatable plants taking over—proving, as Savory observed earlier, that insufficient grazing is just as harmful as overgrazing. Areas with multiple herbivore species had the highest plant diversity and richness at the end of the year, followed by the plots with one species. This, the authors write, shows that a diverse population of herbivores has the power to positively change the composition of the grassland.[18] But that discovery doesn't translate to American grasslands. The authors pointed out the following about their Kansas test site, Konza Prairie Biological Station: "At Konza, although white-tailed deer (*Odocoileus virginianus*) occasionally co-occur in low abundances with bison, there is no true 'multiple herbivore' treatment comparable to that in South Africa."[19] In other words, the Great Plains was never a true multiple large herbivore environment like Africa; buffalo were the keystone species here. In fact, the Kansas plot performed equally well with single and multiple herbivores. So more diversity is preferable, but not required, on the Great Plains.

As far as American ranches go, Great Plains Buffalo is pretty diverse, even if not every species impacts the grassland. Chickens and milk cows provide food for the family, while horses provide transportation over the rough terrain. In terms of large wildlife, mule deer, white-tailed deer, and antelope love the ranch for its rich grass. Phil also keeps cattle, eight hundred cow-calf pairs and one-hundred-some yearlings. Raising two species for profit is already more than most conventional ranchers do. Like farmers, ranchers obeyed calls for specialization under "get big or get out." Many focused on a particular breed or special market area, such as bulls, bred heifers, or slaughter-ready calves. They altered their operations to meet the demands of concentrated animal feeding operations, meat processors, and consumers. Also like farmers, they adopted conventional practices one new implement at a time, creating today's input-heavy meat system that is one of the worst outcomes of industrial agriculture.

Raising buffalo is one way of flouting this system. Unlike cattle, hogs, and chickens, they are a relatively new form of domestic livestock. No scaffold of agribusiness products and services surrounds them—yet. Running buffalo on the Great Plains also represents a physical and psychological return to the native prairie. When we see the buffalo, we remember what the prairie used to be: an intact and thriving ecosystem capable of withstanding the natural wet and dry cycles, an ocean of grass where humans fit into that system instead of disrupting it.

8
The Buffalo

Even though the pickup is dragging a long section of pipeline, the buffalo run toward it when they see us coming. "Normally when I show up they're getting moved. Wherever I go is where they want to go," Phil says. The buffalo are so used to moving that all Phil has to do is open the gate and lead them through with the pickup.

Once they realize they aren't being moved, the buffalo fan out on the hillside to graze. They look scraggily; it's that awkward time in the spring when they don't have sleek summer coats and are still shedding their thick winter hair. Bits of it are half-detached, clinging to their skin. Some are young, some are old. Most have curved black horns. The calves are tan, while the mothers are chocolate brown with camel-like humps. The hair around the cows' teats is wet and matted where the calves have been nursing. Everyone has beards and long skinny legs. They look sort of like African wildebeests, exotic and wild. I'm so unused to seeing the prairie's original grazer, the former keystone species, that they look foreign.

The Jerdes bought their first buffalo cows in 1999. Why buffalo? I ask. "Because they are just so cool," Phil says with a grin. At the time buffalo were popular, a trend that offered the potential for high profits, and many ranchers jumped into the market. "They were expensive," Phil admits. And then the market crashed. "Within about three years the calves were worth fifty dollars. It was a wreck. But we had some help; the folks helped us out. And then we hit the drought about the same time." He's referring to the drought of 2002–6. "So we had a lot of animals that weren't worth anything and not much grass to feed them. It was kind of a perfect storm. But now they're worth some money, which is a good thing."

Buffalo meat is naturally lean, containing less fat than other red meat, white meat, and salmon. An average 3.5-ounce serving of cooked bison has 2.42 grams of fat. Compare that to the same size serving of choice beef at 18.54 grams of fat, select beef at 8.09 grams, pork at 9.21 grams, skinless chicken at 7.41 grams, and sockeye salmon at 6.69.[1] Phil is a proponent of "grass fat," or eating the meat of grass-fed animals rather than grain-fed animals. Grass fat, he says, is good for the body, and he is dismayed that consumers sometimes write off meat altogether because of health effects—effects that he argues come from grain fat put on in feedlots, not grass fat. "If you are gonna eat feedlot fat, then yeah, eat a low-fat diet," Phil says. "But grass fat is where it's at."

It's true that meat from grain-fed animals contains higher-than-ideal fat levels that are neither good for the body nor tasty. Dan Barber, a chef as well as a writer, describes the case of American grain-fed lamb, which arrives in the kitchen with a fat cap an inch or more thick.[2] At one restaurant, Barber cleaned forty racks of lamb a day, throwing 10 percent of each cut away. The French butcher Barber worked under said so much fat was disgusting. That kind of dull, fatty meat would never appear on a French table. As Barber points out, "Feeding an herbivore grain (intensively anyway) is a recent invention, and despite the fact that the practice has become so ubiquitous—and in the case of Colorado lamb, so coveted—it's not actually delicious."[3] Under the concentrated animal feeding operation (CAFO) system, we've come to equate flabby, greasy meat with delicious meat.

Meat from grass-fed animals, on the other hand, contains more heart-healthy omega-3 fatty acids, more conjugated linoleic acid or CLA (a type of fat that reduces heart disease and cancer), more antioxidant vitamins like vitamin E, and less total fat and calories than its corn-fed counterpart.[4] Grass-fed livestock that never see the inside of a CAFO can be part of a healthy diet, not to mention part of healthy grassland. I should say that Phil, Jill, and their children eat this kind of meat almost every day and are in tip-top physical shape. So are their animals. When the soil improves, grass nutrition improves. When grass nutrition improves, animal health improves, meaning animals fend off sickness and put on weight more effectively. The healthier the animal, the healthier the person who eats it.

"I think all the disease is nutrition, or lack thereof," Phil says. "We do purchase some salt and minerals for the livestock. Back during the huge herds of buffalo, they would pass through areas that would be deficient in selenium and then make up for it by going to areas that had too much selenium, for example. Once we put all these fences in and limit the movement of livestock, it's bound to create shortages of certain minerals, which is going to lead to consequences for herd health. So we still are supplementing some minerals, and will be." Minerals are Phil's first line of defense against sickness. He doesn't administer preventative vaccinations, and if a buffalo gets sick, he doesn't turn to conventional medicine. "Actually, the only shot we give is the brucellosis Bangs vaccination and that's more a matter of if we want to sell them out of state; otherwise you can't sell them out of state to a lot of states," he says.[5]

Some might cringe at Phil's philosophy, seeing his approach as cruel because he doesn't step in right away to help sick animals. He explains his view through an example: he and a friend were running cows in the same pasture. The friend's cattle developed pinkeye, and a few of Phil's cows got it, too. Phil didn't "monkey" with his cattle, a rancher's term for "mess with," but his friend doctored his animals with antibiotics. The friend spent a day rounding up the cattle, running them through a chute, and administering the vaccine, a process that caused further stress on the cows. Phil's cows healed up in ten days. The friend's cows healed up in a week—a negligible time difference to Phil. "You can either put antibiotics in the animal and have them heal up in seven days, or not and have them heal up in ten days," he says. "Why do we bother doing a lot of the things that we do?"

He offers another example. He once bought some buffalo heifers and put them in with his herd. The new heifers started dying. Turns out they were full of worms. Phil asked the rancher who sold him the heifers how they had been raised, and he learned they had been living in a big pasture all summer—conventional grazing. Keeping livestock in the same pasture for too long means they are more likely to contract worms because they are exposed to multiple stages of the worm cycle, as Phil explains. "The egg is pooped out, the worm hatches, crawls up the plant leaf, and is there for the next animal to eat and start the process again. What we're doing [on our ranch] is, if that's pooped out, there isn't an animal to consume the next stage of the worm issue. In that instance, we brought those animals back in and wormed them, but it's not a practice we would employ because it's not necessary under our management."

Still, some animals get sick and die. The hard truth is that a ranch is like the natural world: there is death from weather, predators, diseases, and old age. Because of their diet, Phil's livestock are healthy enough to fend off most ailments. Some can't, but that doesn't mean Phil is doing something wrong. No rancher can prevent all deaths, no matter how many inputs he or she uses. "Nature's model isn't that 100 percent of everything lives," Phil says. "So do animals die in our system? Yes. But they only die once. It's just a one-time deal."

In other words, death by a cause conventional medicines *might* have prevented is better than life in the conventional system because it leads, for most livestock, to a CAFO—a place where death is not "a one-time deal," but arguably occurs little by little over many torturous months.

We can see a broad picture of the CAFO system by looking at the life of a typical conventional beef steer. The steer is usually conceived on a cow-calf operation like my parents' ranch, the industry term for ranches stocked with breeding cows that give birth to calves each year. The rancher turns the cows and bulls out to pasture for the grazing season, and healthy cows come back pregnant. Some ranchers artificially inseminate the cows instead. Whatever the case, our steer is born about 280 days later. Cows raise their calves on pasture until they are about six to ten months old, when the rancher weans them away from their mothers.[6] This is the last time our steer will eat grass.

Then he goes to a backgrounding pen on the ranch, a minifeedlot, where he learns to eat corn. He is a ruminant animal, meaning he's designed to eat and digest grass.[7] But steers fatten quickly on grain, and America's corn and soybean growers are producing a surplus of both, making these grains relatively cheap. Ranchers also feed inexpensive, nutrition-poor, starch-heavy ethanol by-products called distiller's dried grains. Many ranchers background because CAFOs don't want calves fresh from the pasture—they would rather let the rancher spend time and money getting the calves closer to slaughter weight, a move that reduces the CAFO's risk and input costs and puts the burden on the rancher. This means many ranchers have become feedlot operators, a way of tailoring their practices to fit industrial standards.

Like my dad. He backgrounds his calves for a few months before taking them to auction. That's a fact I hate revealing to people: we run a feedlot part of the year, which sounds (and is) much less noble than simply being a cow-calf operation. He feeds the calves a mixture of corn silage, ground hay, distiller's dried grains, low-level antibiotics, and corn, just what CAFO owners want him to do. The calves require feed daily, meaning he can't leave the ranch for more than twenty-four hours from November to February and burns expensive fuel to power the feeding equipment. He also grows GM corn that becomes silage and feed grain. Not all ranchers background, of course. The fact that my family does, though, means the meat we eat from the ranch is little better than CAFO-produced meat in the grocery store, because the calves share basically the same diet.

Let's assume our steer is a good-sized animal after several months in the backgrounding pen. The rancher sells him with the rest of the calf crop, sometimes at auction, sometimes to a private buyer. Our steer will likely end up at a CAFO whose mission is to fatten animals for slaughter "efficiently" by keeping them in grassless feedlots and feeding them carefully controlled rations—a processing called finishing. The larger the CAFO, the thinking goes, the more efficient it is. About half of the cattle slaughtered in the United States come from just twenty giant CAFOs.[8] America's CAFO system of meat production has been called the "industrial-grain-livestock complex" because the government indirectly subsidizes industrial livestock operations through the subsidization of grain, which gives meat an artificially low price in the grocery store and encourages grain-fed, not grass-fed, meat.[9] The environmental, social, ethical, and sustainability problems with CAFOs are thoroughly documented (see the film *Food, Inc.* and the books *Fast Food Nation* and *CAFO: The Tragedy of Industrial Animal Factories*, for example), so I'll only mention the most troubling realities our steer is bound to endure.

First, he will eat a diet of corn, soybeans, milled grains, corn silage, maybe ethanol plant by-products, and other fillers like ground-up chicken or pig parts. These rations are usually laced with synthetic growth hormones, so animals gain weight faster, and antibiotics to preempt the diseases that come with standing and sleeping in their own manure 24/7.[10] And it takes *a lot* of antibiotics and hormones. Of the antibiotics sold in the United States, about 80 percent are fed directly to livestock like chickens, hogs, and cattle.[11] Growth hormones are banned in poultry and swine production, but their use remains legal in dairy and feeder cattle. About 80 percent of U.S. beef cattle are injected with synthetic growth hormones.[12] CAFO advocates say this is efficiency at its finest—more meat in less time at a lower cost, with help from science. We used to fatten cattle slowly on grass, maybe with a little grain near the end, and slaughter them at four or five years old. Now we've sped the fattening up to fourteen months, sometimes less, and cut the grass out completely.[13]

The danger of growth hormones lurks under the surface, in some cases underwater. Researchers confirmed that hormones are routinely present in waters near CAFOs and that these hormones disrupt the reproductive behavior of fish.[14] Scientists are researching whether hormone residues in meat and drinking water could cause similar hormone disruption in humans, which could lead to reproductive issues, developmental problems, and cancers already linked to hormone imbalances such as breast and prostate cancer. Happily, our steer has no knowledge of these facts, and if he did, he probably wouldn't care because, let's be honest, he's got bigger problems, such as being stuck in the hell-on-earth that is a CAFO.

The constant feeding of antibiotics at low levels on such a large scale for the past sixty or so years is, as scientists confirm, encouraging antibiotic-resistant bacteria that's showing up in human bodies. We've compromised humankind's ability to fight infections because we consume antibiotics in almost every meal that includes beef, pork, chicken, or dairy. Scientists say it's possible that antibiotics, the twentieth-century wonder drugs, will become useless in this century. The Obama administration tried in 2013 to curb use by banning antibiotics for the purpose of making livestock grow faster (one side effect of prolonged, low-level antibiotics is that animals put on weight). The administration also required producers to get a prescription for antibiotics from a veterinarian before treating sick animals. But lawmakers left a giant loophole: producers can simply argue that they use antibiotics to *prevent* sickness, which doesn't require a prescription, meaning the situation is essentially unchanged.[15] People in other countries are understandably terrified at the prospect of losing antibiotics. The European Union in 2006 fully banned their use in animal feed, though some of its member countries had antibiotic bans in place for decades prior. Nor does the EU want its citizens ingesting synthetic growth hormones in meat or water, so the use of hormones is also illegal. Since 1989 the EU has banned the import of beef containing growth hormones, which is mainly U.S. beef.

Hormones and antibiotics aren't the only problems. Of the pesticide residues North Americans consume, 55 percent originate from meat and 23 percent from dairy.[16] But how can livestock-derived food contain pesticides, which are sprayed on field crops? The answer lies in the conventional grain CAFOs feed to the animals. About 80 percent to 90 percent of the CAFO diet is corn.[17] This grain is chock-full of pesticide residue, and as animals eat the grain the leftover chemicals accrue in their flesh. Considering that meat consumption has risen dramatically in the U.S. over the last one hundred years, and the use of pesticides has increased along with it, it's safe to say that our bodies are more toxic than ever. So are the cattle we're eating. Grain-heavy diets make cattle sick because their rumens stop functioning, causing bloat. Their rumens also acidify and develop ulcers. The biggest issue is usually the liver, which develops abscesses and shuts down. This is why cattle only stay in the CAFO for a few months—they would die if forced to live on corn much longer.[18]

So we're eating sick, pesticide- and hormone-filled animals and expecting to stay healthy. The absurdity of that thinking isn't lost on Phil. "We're created to be healthy; we're not created to be sick. When we are fed things that keep us healthy, we tend to be healthy. Same is true of our animals," Phil says. "But if we are going to go eat McDonald's and drink liters of soda pop, you can expect to need a lot of healthcare. The same is true of our animals. If we are going to feed them distiller's waste, we can expect to reap what we sow. We can expect some problems healthwise with that. When we load up a rumen, which is made to process grasses, on starch, we can expect problems with that."

Phil has a story that confirms how sick these feedlot animals actually are. Several years ago, during the winter and not terribly far from the Jerde ranch, a semi-truck carrying feeder calves tipped over on a curvy, narrow road. Miraculously most of the calves lived, but a few died. One body was left behind, having fallen too deep into a gully to be retrieved. In western South Dakota there is much talk about the supposedly out-of-control coyotes, which presumably would have devoured a dead calf during the cold winter. "Up in the hills, with all these coyotes, nothing would touch that animal," Phil says. "Nothing would eat it. That's a finished grain-fed beef coming out of Canada to be processed in the States. Even our scavengers won't eat it. That's what is being fed to us as USDA Choice. I should have taken a picture because it stayed there all winter long. What can they sense that we don't? And we're the advanced ones supposedly. We'll lose a buffalo from time to time and you come back a week later and it's stripped down to bones, it's spread out."

Where does Great Plains Buffalo fit into this CAFO-driven meat production system? Phil raises buffalo, not cattle, so his situation is different—but not so different that industrial processes haven't been adapted for buffalo. Of the roughly sixty-one thousand buffalo slaughtered each year, the majority are grain-finished in feedlots, a reflection of the agribusiness determination to stick with the grain model and even apply it to a wild animal.[19] It's also a reflection of consumers, who want buffalo to taste and look like corn-fed beef. Buffalo meat is by no means gamey, but it is leaner. Consumers who see thick marbling and fat caps as signs of quality are often disappointed with buffalo, interpreting its meaty taste as inferior.

Even the anti-CAFO Phil sometimes has to sell some animals to feedlot operators because they are the major buyers. Most of the time, though, he sells live animals to people wanting to finish them on grass and eventually slaughter them. That way he participates in the regenerative meat system. He's also sold live cattle and buffalo to people who want to start their own ranches. But he can't always find such buyers. "If I could never send another animal to a feedlot, I'd be there," Phil says. "Right now I'm not there. I got mortgages to pay and whatnot, so sometimes animals get sold that way."

To Phil, selling live animals is good, but selling meat would be better. "Ideally I would like to produce grass-fattened food, grass-fattened animals, right off the place and market directly," he says. The problem is that there's no one around to buy it. Remember, the ranch is hours from the nearest urban center. "I don't know how you get around it out here," he says. "There just isn't a large population of people. We've played around with marketing some of our own buffalo, and grass-fed beef for that matter, but getting around transportation costs is just a huge deal.

Right now we're just shipping the animals off and letting someone else take that enterprise on who knows what they're doing."

Phil faces an even larger hurdle, though: the conventional slaughter system. It's illegal to slaughter animals on the ranch and retail the meat. Producers can only sell meat processed in a USDA-inspected slaughterhouse, which is usually far away, and very few slaughterhouses are small enough to accept animals from individual ranchers anyway. Like CAFOs, the megaslaughterhouses that process our meat have a hand in the market—a heavier hand than most of us realize.

9

The End of the CAFO

Let us return to our steer, who's still in the concentrated animal feeding operation, dipping his mouth into a trough of corn. Let's say he's black in color, like the steers that come off my family's ranch. Let's say he has brown eyes, long eyelashes, and a white blaze under his belly, like one of my best steer friends from childhood, a puppylike boy named Bruiser. Let's also say he is ready to be harvested, having reached the conventional slaughter weight for steers of twelve hundred to fourteen hundred pounds, which took four to six months of finishing at the CAFO.

He trots into a semi-truck and heads to the slaughterhouse, where a man holding a pneumatic device called a stunner approaches on a catwalk. The man rests the stunner, or knocker, between our steer's eyes, and a metal bolt shoots out, enters his skull, and pierces his brain, killing him. Roughly thirty-one million cattle faced the knocker in 2016.[1] The USDA reports that 814 slaughterhouses were in operation under federal inspection that year, which sounds like quite a few until one reads that the thirteen largest plants slaughtered 55 percent of the cattle and 60 percent of the hogs, and the largest three slaughtered 54 percent of the sheep and lambs.[2] The biggest processors in each category often overlap. Some of these behemoths slaughter five thousand cattle a day or more.

But here's the deceptive part of the numbers: those thirteen plants aren't individually owned. It's thirteen major *locations*, but only a handful of companies own them. Tyson, Cargill, JBS, and Leucadia together control almost 75 percent of all beef slaughtered in the U.S.[3] Similar monopolistic conditions exist in the slaughter of pigs and chickens.

The consolidation of slaughterhouses into fewer hands means America's cow-calf producers, like Phil, are stuck. Most of today's slaughterhouses (and all of the megaslaughterhouses) refuse to buy and slaughter cattle directly from ranchers. It's too much of a hassle to work with individuals. They only buy livestock in bulk from CAFOs—and in many cases, the slaughterhouse owns the CAFO, or owns the subsidiary company that owns the CAFO. Small plants will deal with ranchers, but these are dwindling in number. Even if ranchers could take animals directly to a big slaughterhouse, there's no way they could get their meat back. Slaughterhouses mix all the meat together and slap on the company label, making it impossible to identify what cut of steak or pound of ground beef came from which animal.

It wasn't always this way. Small slaughterhouses used to be the norm in America, at least before the major Chicago beef packers rose to power in the 1900s—Big Slaughter. Most slaughterhouses in the pre-Chicago days were family-owned establishments that processed local or regional animals. While they differed in size, most didn't slaughter thousands of cattle a day. When "get big or get out" became the mantra of agriculture, though, slaughterhouses consolidated even more, big ones pushing smaller ones out of businesses until only a few massive plants remained. Today's slaughterhouses pay mostly foreign workers a pittance to labor under hellish conditions and provide little or no help if they get hurt or killed (which frequently happens in the meat-cutting business).[4] It's much like the life migrant farmworkers endure, but behind closed doors that journalists are usually not allowed to enter.

If that doesn't turn your stomach, maybe this will: when our steer arrives at the slaughterhouse, he's caked with manure. He's been standing and sleeping in it for months. After he's dead, workers peel away his manure-crusted hide, trying to prevent manure from touching the meat but not always succeeding. That manure contains deadly microbes, one of the most lethal being *E. coli* O157:H7. Eating just ten of these bacteria can kill a person.[5] *E. coli* O157:H7 evolved in feedlots, in the rumens of cattle like our steer. The rumen is not an acidic place when cattle eat grass, but it acidifies when they eat corn. Inside the fermentation tank that is the rumen, *E. coli* O157:H7 developed resistance to acid, which means it doesn't die as many bacteria do when they enter the acidic human stomach.[6]

That's what drops onto our meat at the slaughterhouse. Instead of switching cattle to grass and keeping them in pastures, which would prevent most *E. coli* O157:H7 and manure-caked hides, the slaughterhouse kills bacteria by using irradiation, chemical sprays, and steam. This isn't fixing the problem; it's applying a bandage. The bigger problem is that, if meat does become contaminated, the bacteria end up in countless packages, untraceable. Because our meat comes from a handful of plants, one instance of contamination can spread nationwide through the sprawling food chain, just as contaminated greens do.[7]

Megaslaughterhouses and CAFOs are what often convince people to become vegetarians. The conventional system makes even me, a rancher's daughter, question the ethics of eating animals—until I remember that it doesn't have to be this way, and has only been this way for the last half-century or so. There's actually beauty in consuming animals because it represents what I think is a symbiotic relationship not unlike the symbiosis between the buffalo, birds, and insects. When we raise animals with care, offering them a dignified life, we show respect for beings that will ultimately give their lives for us. We meet their needs and they meet ours, a symbiotic relationship. CAFOs and megaprocessors take away the beauty of that connection. The animals get very little in return for their sacrifice, certainly not a dignified life. We get our meat, but with environmental, health, and social costs that are far higher than we can bear.

Perhaps organic certification could take care of some of these environmental and ethical concerns. Great Plains Buffalo was certified organic by the USDA from 2005 to 2009, but Phil stopped participating in the program for three reasons, the first being, quite simply, that he wasn't getting paid any more for certified organic live buffalo than grass-fed live buffalo. Why not remain certified anyway, I press on, just in case prices change? "There came a requirement that to be certified, through the certifier I was with, that we had to grant access to any part of our property, at any time, to inspection. Without notice," Phil says indignantly. "I'm fine with them inspecting the property, but I want to be notified ahead of time that 'We want to come and look, will you be around?' I'm not just going to turn somebody loose anytime they want to go into any pasture because there might be a thousand buffalo in the pasture they're going into." People who aren't used to buffalo could trigger a stampede, causing them to escape and damage neighbors' property. They might get too close to a calf, prompting a mother cow to charge. A territorial bull can trample and kill a person in seconds. If something like that happened, Phil would be liable, a risk he wasn't willing to take.

Then came a watering down of the organic rules, the main reason Phil decided not to renew his certification. "The government got involved in making the rules, and all the sudden I could pour Ivomec on my cattle as long as I had a good reason, and they could remain certified," he says. Phil is right: the organic rules have significant loopholes. Parasiticides like Ivomec are allowed in breeding stock if preventative measures fail, and as long as the cow isn't nursing or in the latter third of a pregnancy. The USDA also allows the use of almost all vaccines, which it categorizes in a single group rather than differentiating between different drugs. "To me, once the government got involved with it, the terminology lost its meaning," Phil says. "'USDA organic' is still fooling people, but more and more people aren't being fooled. With this local food push—which isn't a big deal here, but it is in bigger metropolitan areas—of knowing where the livestock came from, knowing where the food is grown, those people aren't going to be fooled so easily. I don't see us going back and getting re-certified again."

I see Phil's point. The USDA National Organic Program (NOP), which determines organic standards, controls and enforces organic food regulation, from the smallest one-acre farms up to the largest multi-ingredient food processors. Perspectives on the strength of the NOP vary depending on who one talks to or reads. Many critics argue the standards are too lax as a result of agribusiness influence, that the voices of organic growers were lost in the codification process of the late 1990s. These growers saw organic as a philosophy much like holistic management: the farm is part of an ecological whole that works with nature rather than against it. The NOP rules fail to capture that sentiment. In *Rebels for the Soil*, Matthew Reed claims that "in being seen as just another 'brand,' the wider ecological and social messages of the organic movement have been lost."[8] So Phil's right to be concerned that the little green USDA certified organic label doesn't necessarily indicate a sustainably grown product.

Reed also argues that the NOP is inefficient and mired in scandal, a view other researchers corroborate.[9] The Cornucopia Institute, a nonprofit research organization based in Wisconsin, calls the current situation the "Organic Watergate" because of the "cozy relationship between the USDA and agribusiness lobbyists," typified by the USDA's consistent placement of agribusiness representatives on the National Organic Standards Board (NOSB), a board that is intended to be a watchdog of the NOP.[10] They cite, for example, the slot reserved for a scientist being filled by a representative from General Mills, who originally was appointed to the consumer/public slot but was moved due to public outcry. The environmentalist position was once filled by a representative from Purepak Inc., who had no background or degree in environmental issues. These appointments matter because the NOSB actually has teeth; it must review and approve any input or synthetic ingredient before it becomes legal for use in organic food production. Critics say board members do not separate the interests of their agribusiness and food retail employers from their decisions on matters of organic food policy.

Plus, the board receives its information secondhand from a separate technical advisory panel that provides scientific reviews of synthetic materials. The Cornucopia Institute alleges that these reviews are biased and flawed, and part of the problem in identifying misinformation stems from the fact that the parties conducting the reviews are no longer disclosed.[11] The institute looked into scientific reviews from the past, when the reviewers were named, and they found a disturbing trend: "What we found is that past technical reviews have generally been produced by corporate executives, consultants serving corporate agribusiness or closely aligned academics. Many of these technical reviews have grossly downplayed health and environmental risks associated with petitioned synthetic materials."[12] And these violations occurred *before* a veil of secrecy protected the identities of the scientific reviewers. It's difficult to argue that secrecy has resulted in more accurate, less-biased science in this case.

What's more, the large-scale farms producing most of the nation's organic, supposedly sustainable food actually aren't that different from conventional farms. They obey organic rules—but barely. This is known as "shallow organic" or "industrial organic." These farms grow crops in large monocultures, treat symptoms with allowable chemicals instead of using regenerative farming techniques, and don't restore soil naturally. The livestock equivalent is the organic CAFO or organic poultry barn that holds thousands of chickens. The food coming from farms and ranches practicing shallow organic is hardly better than conventional food in a nutritional and environmental sense, because these farms ignore plant, soil, and animal health. They claim to be sustainable, but in reality they are not. Perhaps some producers out there share a common conception of what makes food organic and sustainable, and perhaps they would follow unwritten standards if left alone, like Phil does. But in terms of the national organic food supply, we're not dealing with individual farmers and ranchers—we're talking about giant agribusiness corporations eager to profit from the growing consumer demand for organic food, corporations that have already shown resistance to strengthening organic rules and try every year to weaken them.

Super-sized corporations like Kraft, Nestlé, Coca-Cola, Dean Foods, and General Mills produce the majority of the processed food labeled USDA Certified Organic.[13] Many of the world's major food processors have shifted to an "if you can't beat them, buy them" attitude by acquiring top organic brands while continuing to profit from nonorganic sales, and they fight measures that would encroach on those profits. General Mills, for example, owns the Muir Glen and Cascadian Farms organic food brands, but the company joined a consortium with Nestlé USA, PepsiCo, Monsanto, and other food giants that together spent about $22 million on advertising and lobbying to kill bills in California and Washington State that would have made the labeling of GMOs mandatory.[14] Behavior like this makes it clear that those who own most of the organic food industry don't actually believe in the organic philosophy—the sustainability they are selling is fake.

Remember our conventional steer, the black one with the long eyelashes and the white blaze under his belly? Let's pretend he's an organic animal instead. Doing this will help us see where current organic standards succeed or fail. He is still conceived and raised on a cow-calf operation and weaned at the same age. He spends a minimum of 120 days a year out to pasture, depending on the length of his region's grazing season. He lives on pasture year-round if weather and grazing conditions permit, and if they do not, then he at least has access to the outdoors year-round. His pasture and any supplemental feed are certified organic, except for trace minerals and vitamins, and a minimum of 30 percent of his dry matter intake (DMI) comes from pasture. Sounds good so far.

He is not fed growth hormones, low-level antibiotics, or other prohibited substances (urea, arsenic, and slaughter by-products are just a handful of such banned substances that CAFOs routinely feed). To keep him from falling ill, the rancher uses preventative management strategies—disrupting the worm cycle like Phil does, for example—but if he does become sick and approved substances (which may or may not be truly organic) fail to cure him, the rancher can let natural selection take its course or use unapproved medications, like antibiotics. If the latter happens, he is no longer considered organic and will have to be sold into the conventional market. Sounds fair. For our purposes, though, let's assume he remains healthy, since pasture-raised animals generally stay healthy anyway.

Our organic steer might be finished on grass and sent directly to the slaughterhouse—that would be ideal. Instead, there's a good chance someone will finish him in a CAFO. He is exempt from the 30 percent DMI pasture requirement for the last one-fifth of his life, up to 120 days. That means he can still be finished on corn. The corn is certified organic and non-GM, meaning pesticides won't linger in his meat. But he'll still suffer all the ill health effects of a corn-heavy diet and live in a sea of manure. This is a major problem with the organic standards: they don't fix the CAFO issue. Many organic supporters would argue that CAFOs conflict with the overall organic philosophy of health, sustainability, and respect for animals.

On a brighter note, the slaughterhouse that awaits our steer will be certified organic, which according to the USDA means that (1) the premises are cleaned with organic substances; (2) the processors do not comingle organic and conventional meat; (3) if the steer is kept on the premises, then he will be fed organic feed, but the goal is to keep him on site for no more than half a day; (4) the slaughter methods minimize fear and stress; (5) the plant continues his individual record started by the producer; and (6) the packaging products are free of synthetic fungicides, preservatives, or fumigants. A good start. But the meat from our steer will still be mixed with that of many other organic animals, and there's still the risk of *E. coli* O157:H7 and other bacteria contaminating it. The organic slaughter system is much the same as the conventional system, with a few cosmetic changes.[15]

The real problem here is the CAFO. Even conventional grass-fed beef is better for the environment, our health, and the animal than organic feedlot meat. The ideal is organic *and* grass-fed, which produces 40 percent less greenhouse gas and requires 85 percent less energy (i.e., corn that requires fossil fuels to produce) from birth to slaughter than CAFO-raised meat.[16] Raising animals on pasture means their waste becomes valuable energy in the soil instead of toxic sludge in lagoons. Grass is actually as good or better than corn at putting pounds on animals. "You put cattle out to grass in the springtime, early summer, and you get two-and-a-half- to three-pound-a-day gains. That's feedlot type of gains," Phil says. "So why don't people get gains like that for ten months out of the year here? We don't because we don't have the quality forage. But if we could have leafy green forage the whole growing season, all the sudden instead of me moving buffalo around, just grass-raising them, all the sudden that's a minifeedlot if I can produce the gains and do it on grass."

Proponents of the CAFO system argue that cattle can't get fat enough for slaughter on grass alone—they need grain. Given the low nutritional state of America's ranges, ranchers don't see how cattle could achieve the type of weight gain needed for slaughter. If ranchers switched to grass-finishing cattle today, the effort would likely fail—unless they also altered their management style. Pastures managed holistically like Phil's contain more than enough nutrition to prep cattle for slaughter. Recognizing that quality forage is more efficient than feedlot rations could revolutionize the cattle industry. We could ditch the CAFO entirely. Of course, ranchers would have to adopt regenerative agriculture, and consumers would have to prioritize grass-fed meat. But the potential is there. As Phil puts it, "It could be a revolution of the feedlot mindset."

10

The Sun's Wealth

Phil's son Viggo, two, who we exchanged for Quilla at the house earlier, naps on the pickup seat, his head in his brother's lap. We bounce across the prairie—sometimes we hit ruts so deep that my rear end leaves the seat—but Viggo sleeps for an hour or so. "Nothing like a bumpy ride across the prairie for napping," Phil says. Indeed.

I remember pickup rides just like this one with my dad, mom, and siblings. On Sundays we would climb into the 1970s-era red and black crew cab Ford, another hand-me-down from my grandparents, and drive salt blocks and sacks of loose mineral to the cows. My siblings and I would wet our fingers and dip them into the mineral so it stuck, then lick the powder off. It tasted salty and metallic. Back then we had 150 or so cows and kept them in one pasture all summer, which we creatively called the summer pasture. We still call it that, even though it's too small for the herd now and serves instead as a temporary pasture where they spend part of the fall before coming home for winter. We always hoped the cows were on the other side of a steep draw so we could gasp and cringe as Dad went off-roading. The cows gathered around the pickup as he filled the salt and mineral tubs, and by hand we fed the friendly ones cake, tasty pellets as round as cigars. When Dad parked the pickup truck by the stock dam, we leapt out and followed cow trails deep into the cottonwoods, tasted ruby-red buffalo berries and wild purple plums, smelled sunflowers and sweet clover.

Sometimes Dad would bring a long, black rubber tube filled with fly powder, which he slung from two poles so the cows could walk under and rub their backs on it, shaking the powder loose. Dr. Scratch, the device was called. When summer turned to fall, we brought creep feeders, boxy structures filled with pellets dispensed from two troughs that we pulled behind the pickup and left in the pasture, refilling as needed. Dad wanted to get a head start on fattening the calves before they got to the backgrounding pen. I loved climbing up the ladder and standing on top of the feeder, looking at the prairie that stretched forever. If I stood very still, a black calf would wander to the trough and start eating. When I went home a few months ago, Dad said he adds low-level antibiotics to the creep feeders now.

It's amazing what we see when we reexamine our childhood. In college, when I would conjure childhood memories like these to write about in nonfiction workshops, I used to see the friendly faces of cows, the yellow coneflowers I picked for my mom, the blue, long-sleeved checkered shirts my father wore during the summer. When I look at the same memories now, I mostly see the parasiticide powder, the calves eating pellets instead of grass, the cattle scattered across the land. Conventional agriculture hasn't just ruined the farm for me as an adult. It's ruined huge chunks of my childhood as well.

It's unlikely that will happen to little Viggo, his mouth dropping open as he sleeps. I envy him a little. When I retrace my family's journey on the industrial path, I always come back to "get big or get out." Since that infamous threat from Butz, the government has encouraged farms to get bigger and adopt conventional practices through tax incentives, subsidies, and other federal programs, weeding out small farms in the process. "Government involvement in agriculture has picked the winners and really picked the losers, too," Phil says. I ask him to explain what he means and, in typical Phil fashion, he turns to a book: *The Tipping Point* by Malcolm Gladwell.

"The most efficient-sized business, it's kind of a family-sized business, maybe a few employees. But once you get past a certain number of employees your efficiency goes down," he says. Phil can't remember the number offhand, but when I read Gladwell later, I learn Phil is referring to the Rule of 150. This rule says that roughly 150 is the maximum number of people we can have a real social relationship with. If groups, organizations, or businesses get bigger than that, it's best to divide. This is what the Hutterites, a religious group somewhat similar to the Amish, do: when their colonies reach 150 people, they split up. At Gore Associates, a multimillion-dollar tech firm in Delaware profiled by Gladwell, manufacturing plants have no more than 150 people, even if that means building fifteen plants within a twenty-five-mile radius, as the company has done. Expanding a business beyond 150 people is a little thing that makes a big difference in productivity, worker morale, efficiency, and overall success.[1]

Phil brings up the Rule of 150 not to say that farms and ranches should have 150 employees—that would mean the operation is huge and industrial-sized—but to show that efficiency does not go hand in hand with size. Bigger is not necessarily more efficient. So why, he asks, have farms ballooned in size if efficiency is lost in the process? He answers his own question: "I would argue it's because of the rules set by government that favor the large players over the little ones. Once you set the rules that favor the big guys, why should we be surprised that that's what we get?"

When Phil speaks of the "rules that favor the big guys," he means the shaping of government policy by massive private agribusiness corporations to their benefit, and to the benefit of large farms that purchase their products. In *Let Them Eat Junk*, Robert Albritton outlines three reasons for this corporate shaping.[2] First, people sympathetic to the corporations they are tasked with overseeing often control regulatory agencies. I hear a similar critique from Phil: "Have you ever seen the revolving door of chemical companies, Monsanto and whatnot, to government and back and forth?" he asks, disgusted. As Albritton notes, "In the case of GM seeds, they were approved by the U.S. government based entirely on research funded by the corporations who wanted to market the seeds. There was no independent government testing."[3] What oversight committee, except one composed of corporate executives standing to make money, would allow such one-sided research to prove the safety of a new product?

Second, Albritton points to costly election campaigns that force politicians to seek corporate money, making them beholden to corporate will. Agribusiness corporations are heavily involved in elections, particularly in rural states, and they are rewarded for their campaign contributions with friendly legislation. Third, corporate lobbies not only influence congressional members, but also write the legislation, which our representatives turn into law. Albritton's fourth point is that corporations systematically silence their opponents, going after critics, media and farmer alike, with lawsuits. Monsanto is notorious for this. Corporations have also tainted agricultural research by funding studies with their own "organized groups of scientists and think tanks to influence public opinion and to provide Congress with supposedly 'authoritative' information which advances their short-term corporate profits even when the long-term social costs may be immense."[4] For independent researchers, coming up with the "wrong" scientific conclusion about seeds, chemicals, or fertilizers can be the end of their career, just as we saw in the early days of climate-science research that was at odds with oil companies.

All of this might appear too bizarre to be true. Could agribusiness companies really have created, as *Food, Inc.* director Robert Kenner puts it, "an Orwellian world of behind-the-scenes wire-pullers controlling a fundamental aspect of our lives—the food we eat—which operates in near total secrecy thanks to the fear it instills in people who know about it"?[5] Well, yes. Consider today's food libel laws, sometimes referred to as veggie libel laws. Thirteen states have food disparagement laws that make criticizing food products illegal, including South Dakota, where I grew up, and Florida, where I'm writing this book.[6] In those states, critiquing food is illegal, even if that critique is fair and truthful. Make no mistake: the agribusiness lobby had a strong hand in creating these laws. If that's not an Orwellian world of behind-the-scenes wire-pulling, I don't know what is.

Even if food libel laws weren't a reality, it's unlikely most people would know what to say about food anyway, because agribusiness companies have been highly successful in preventing food labeling laws. The respected food writer Marion Nestle has written on this topic for decades. What's funny is that corporations often argue that they aren't accountable for the consumer's health and that consumers bear responsibility for their food choices—but these corporations won't give people the information they need to make educated choices in the first place.[7] Genetically modified food isn't labeled. Neither is meat from cloned animals. Today's labels are hardly helpful anyway, written as they are in what seems to be a foreign language. Corn-derived food products often fly by many different names to disguise their corny origins. Maltodextrin, xanthan gum, and sorbitol are just a few examples of ingredients we scratch our heads over—*What the heck is alpha tocopherol?*—that are corn-derived.

Corporate control over agriculture policy has worsened since the *2010 Citizens United v. Federal Election Commission* case freed corporations and unions to spend unlimited amounts of money on "electioneering communications" like TV commercials and mass mailings. Under the ruling, corporations funnel money into political action committees commonly referred to as "super PACs," and these PACs are the official sponsors of the communication. Although direct campaign contributions to candidates are still illegal, we saw in 2012 how quickly these PACs materialized, flooding America's televisions, airwaves, and mailboxes with messaging. Voters heard the voice of big business, including agribusiness that spent about $9.4 million (roughly $7.4 million on Mitt Romney, $2 million on Barack Obama), louder and clearer than ever before.[8] In the 2016 election, the agribusiness sector donated less, roughly $2 million to Hillary Clinton and $2 million to Donald Trump.[9]

The problem isn't voicing political messages. It's how much control the promoters of these messages have over elected officials once the smoke of the election clears. If a corporation heavily supports a candidate, that candidate is beholden to protect and promote the business interests of that corporation, lest they abandon him or her in the next election cycle, launch a smear campaign, and throw company support behind the opposition. Of course, individuals can also support campaigns and super PACs and leverage their interests just like corporations can; think of casino owners Sheldon and Miriam Adelson's combined $30 million donation to Mitt Romney's super PAC or the Koch brothers and their numerous "investments" in public policy. But corporations and individuals are generally not equal in power. A famous (or infamous?) quote comes to mind here: during his 2012 presidential bid, Romney said to one reporter that "Corporations are people, my friend." Supreme Court Justice John Paul Stevens, and countless other philosophers, economists, and ordinary Americans, would likely disagree with that assessment. Stevens claimed in his summary of the court's dissenting opinion on the *Citizens United* case that the majority committed a grave error in viewing corporate speech as equivalent to human speech.[10] It's easy to see why. Aside from the moneyed 1 percent, most Americans do not have the financial or political clout of corporations. Our voices are just not as loud. Unlike the speech of corporations, usually expressed through advertising, exerting pressure on politicians, and filing lawsuits, an individual's speech rarely has the same wide-ranging impact on public opinion, laws, and choices consumers make in the grocery store.

Eric Schlosser predicted in his 2001 book *Fast Food Nation* that the twenty-first century "will no doubt be marked by a struggle to curtail excessive corporate power."[11] He reiterated this belief in the 2012 afterword, citing examples of individuals and groups rejecting and rebelling against corporately controlled food—a good sign, but proof that corporate control is still a major problem.[12] The *Citizens United* ruling represents a reverse in the fight to end corporate control of public policy. It reaffirmed the entrenchment of corporate power in our political system, setting a precedent for future rulings that equate corporations with people or even favor corporate rights over individual rights. The election of Donald Trump, a man who represents big business at its worst and who has filled his cabinet with corporate executives and pro-agribusiness, anti-environment leaders, is also a reverse in this fight. Any headway that has been made against corporate control of agriculture is likely to recede under this administration.

Corporations have one goal: profit. Under conventional agriculture, so do farmers. It's inputs versus outputs—except many farmers receive so much government assistance that the output part of the equation, the profit, isn't accurate. "We measure success by profit, right? Well, if you throw in government payments then that skews what ends up being profit," says Phil. These programs include subsidized crop insurance, payments when commodity prices fall below certain thresholds, aid if crops fail because of weather-related disasters, and low-interest loans. Tax loopholes allow farmers and ranchers to write off equipment purchases, buildings, and other agribusiness expenses. This assistance often hides the truth that farmers are spending more than they're bringing in. So are they really succeeding and being efficient? Phil wonders. He answers his own question again: no.

I ask if Great Plains Buffalo participates in government assistance programs. Phil jokes that they use interstate highways, but that's about it. "We were in the farm program up until about three or four years ago and then we got out of it," he says. "It's actually been a good experience, just not following their rules." Rules are an understatement: participating in government aid programs involves heavy paperwork and regular trips to the local USDA Farm Service Agency office. For some programs, farmers have to plant crops by a certain date to qualify for aid. Phil doesn't like being told what to do, like he didn't appreciate having unannounced inspectors on his land. He abstains from aid programs on principle, to show that farmers and ranchers don't need them.

Phil realizes that conventional farmers often can't make ends meet without government support, so he understands why people enroll and doesn't hold it against them. It's hard to turn down free money when you're trapped in a system that requires you to buy inputs. But when practically everybody receives assistance, it can be tough for people who don't to compete. "We go to a bull sale or a land-lease auction and see who the winning competitive bidders are, and then look back at the farm bill and who's funding them," Phil says. "Using your own tax dollars to bid against you, basically."

Phil would like to see profit defined differently—not in monetary terms, but in environmental terms. "Profit is our measure of success, but what if we chose something else to be our measure, like soil carbon levels?" he says. "We have a history of less and less carbon in our soils. We've farmed that out or grazed it out in exchange for paper dollars and called it a success. It's not a long-term, sustainable practice. Carbon in the soil is what allows soil to have water storage. So when we remove carbon, all the sudden we don't store the water, our droughts get more frequent and more intense. Our wet times lead to flooding and soil erosion that way."

Money won't help us farm forever, but carbon will. So will clean water, symbiotic relationships, and fertile soil. The truly successful farmer isn't the one with the most money, but with the healthiest land. Conventional agriculture doesn't see farming this way, though, and neither does the government, which, as dictated to by agribusiness, essentially pays farmers to continue unsustainable practices. "Those of us on the land, who *are* responsible for desertification, don't realize we are, don't have a stake in it, are being paid not to [stop desertification] through our commodity programs and whatnot, and it's extremely profitable for corn and soybean farmers who are making money like they've never made it before," Phil says. "At the same time, the organic matter has been farmed out of the soil. You have to put all that nitrogen and phosphorus into the soil in order to grow that crop. Once the organic matter was farmed out of the soil, now it's just basically a medium to hold the plant up, and all the nutrients, all the nitrogen, is added in the form of inputs."

Just a medium to hold the plant up—that's exactly what soil is in the conventional model. There's no need to worry about organic matter or carbon when synthetic fertilizers provide artificial nutrients, or to use natural processes to control pests or disease when you can reach for a chemical. Years of these easy fixes have deadened the soil. Phil provides an example, a farmer friend from "East River," the corn- and soybean-growing side of South Dakota east of the Missouri River. The friend grows high-yielding corn, around 180 bushels per acre.[13] Phil once asked his friend how much he might get if he were to stop putting on fertilizer, pesticides, and herbicides. "He said, 'I don't even think we'd get 50.' All the nutrition that's in that corn is what was put on the field. I'm not sure that anhydrous and whatever else they put on is really all that good."[14]

Farms buy energy, that is, fertilizers such as anhydrous ammonia, at a high cost instead of producing it themselves for free, as farmers did before the advent of synthetic, oil-based fertilizer. They essentially dump oil on their fields to make up for the carbon and organic matter that's gone. And they dump on a lot: it takes 230 pounds of nitrogen fertilizer to grow an acre of conventional corn, 50 pounds of which ends up in the environment instead of in the plant.[15] Synthetic nitrogen is the most-used fertilizer. Nitrogen is one of the three macronutrients, along with phosphorus and potassium (abbreviated NPK, respectively), that virtually all plants need for growth. Most fertilizers, even the most basic garden enricher, contain some balance of NPK. Nitrogen, however, is a special case, first because all life depends on its existence, and second because it's not readily available, at least in a form plants and animals can use:

> Nitrogen is extremely plentiful—it makes up nearly 80 percent of the air we breathe. But atmospheric nitrogen (N^2) is joined together in an extremely tight bond that makes it unusable by plants. Plant-available nitrogen, known as nitrate, is actually scarce, and for most of agriculture's 10,000-year-old history, the main challenge was figuring out how to cycle usable nitrogen back into the soil. Farmers

of yore might not have known the chemistry, but they knew that composting crop waste, animal manure, and even human waste led to better harvests.[16]

In nature, plant and animal decay and waste return NPK and other nutrients to the soil. Other nitrogen becomes plant-usable through a process called "fixing," which means that nitrogen atoms split and join with hydrogen atoms. Soil bacteria residing on the roots of legumes—such as alfalfa, soybeans, or peas—fix that nitrogen so more plants can grow the next season. Legumes fixed most of the earth's nitrogen before synthetic fertilizers appeared.[17] Farmers incorporated legumes into their crop rotations to keep their soils nitrogen-rich. This is how soil wealth came from the sun: plants, powered by photosynthesis, provided nitrogen for life.

Not anymore, though. Today, synthetic nitrogen fertilizers have replaced naturally occurring nitrogen, thanks to a German scientist named Fritz Haber who in 1909 discovered how to create plant-available nitrate using the Haber-Bosch process. The Haber-Bosch process involves taking hydrogen gas from oil, coal, or natural gas and combining it with nitrogen gas from the air under extraordinary heat and pressure provided by copious amounts of electricity. Hence, synthetic nitrogen is derived from fossil fuels and fossil fuel–created electricity.

Creating plant-usable nitrogen with fossil fuels represents a major change in the nutrient cycle—or, rather, it made the nutrient cycle irrelevant. As Michael Pollan observed about the creation of synthetic nitrogen, "The basis of soil fertility shifted from a total reliance on the energy of the sun to a new reliance on fossil fuel."[18] Oil also powers the massive equipment found on conventional farms and the creation of pesticides. All told, it takes the equivalent of a quarter to a third of a gallon of oil to grow one bushel of conventional corn, or fifty gallons per acre.[19] Producing a calorie of food requires more than a calorie of fossil-fuel energy, a net loss of energy. As Pollan writes, "From the standpoint of industrial efficiency, it's too bad we can't simply drink the petroleum instead."[20]

Using costly, nonrenewable oil energy to grow food at a net energy loss is inherently unsustainable, because the oil will run out. But using the sun's free, limitless energy preserves and creates wealth in the soil, which would be a more efficient approach to agriculture—a regenerative approach.

Since the first settlers came to the prairie, though, the model has been extraction instead of regeneration. "We know some folks where, when they first came out and dug that virgin soil, there was grass wagon-high," Phil says. "Well, they dug that soil and planted in it, and it wasn't necessarily that hard to grow a crop because what you were doing was transferring that carbon and that energy in the soil to the wheat seed or whatever it was you were producing, like corn. But all it was, was a transfer of wealth. It wasn't making new wealth with the sun necessarily." Making wealth from the sun—making wealth from nature, for free, instead of using wealth from oil. That's what truly regenerative agriculture should do.

Until farmers understand this, though, they're unlikely to stop seeing conventional agriculture as the most efficient way to farm. After all, 180 bushels per acre is much better than 50 if you have chemicals, GM seeds, fertilizers, and equipment to pay for. Money is also the only measure of success in capitalism—and Phil, a self-described capitalist, sympathizes with the farmer's need to make money. Still, he doesn't think money is the right indicator of success for agriculture. "When we use a different measure such as soil carbon," he asks, "are we moving forward or are we going backwards?"

So what direction are we moving when it comes to agriculture, forward or backward? Is forward progress higher yields, better technology, and more control over nature? Or is it our ability to work with nature to produce food? Is progress related to the economic and social health of rural communities, or the financial health of corporations connected to agriculture? And who is the "we" anyway? Is it farmers? Consumers? Human society as a whole? In Phil's estimation, the "we" of the question is mainly the land, and by extension the animals, plants, and people who depend on it—all of us, as one whole, including consumers, in keeping with his holistic approach to grazing. That's one reason he believes consumers can and should help change agriculture, because they have a stake in the system just as much as farmers do. "I think it's going to come from supply and demand, when the consumer demands real food," Phil says. "It isn't going to come by passing a law or making a new farm bill."

Phil expects individuals to take responsibility for their food choices and be educated about how their food is produced. "Right now there are not enough people who care what they put in their body," he says. "A lot of consumers don't even ask what the price is." There's weight behind the words *what the price is*. Phil means the price paid by the land, livestock, water, community, and people who eat the food. Consumers see a dollars-and-cents price on a food item and think that is what it costs, but the hidden costs stretch beyond money. We only have the illusion of cheap food in America: the dollar price might be low, but the social and environmental cost is high.

Until the consumer demands change, capitalism will insist that farmers, meatpackers, and food processors continue their incessant march toward higher profits. Values, ideologies, common sense, consumer and environmental health—none of these concerns matter to the rational capitalist:

> A rational capitalist's loyalty cannot be to the material or ideological qualities of a thing (that is, use-value), but must always be to profit as pure quantity. For example, a truly "rational" capitalist, no matter how religious, would shift production from bibles to pornography purely in response to profit signals. . . . In terms of food, a "rational" capitalist will produce unhealthy food if it is more profitable than healthy food, and will utilize polluting and toxic chemical inputs as long as profits are increased by doing so.[21]

Most conventional farmers and ranchers are rational capitalists. Like Ryan Roth said, you can't love your crop. They have to be rational capitalists to survive—unless we find a way to make it more profitable to farm regeneratively. A possible step is making it unprofitable to continue producing food the way we do now, which consumers can do by not buying conventionally produced food and voting for representatives who will promote regenerative agriculture. I agree with Phil that laws alone can't solve our agricultural problem—but I do think we can take away the laws and programs that create a mandate for conventional farming and the "get big or get out" philosophy.

Farmers and ranchers also have to take responsibility for the environmental destruction caused by conventional agriculture, particularly desertification and climate change. "Over the course of history agriculture has been the biggest polluter, if you will, putting carbon in the air," Phil continues. "We dig up the soil and right away within days you have a huge amount of carbon put into the air. We do the same thing over time when we take productive grasslands and turn them into bare ground. We just do it slowly over time, so we don't notice it. Those of us out on the land aren't taking responsibility for that. The ones in the city are. We are the ones who have the ability to change it and capture carbon, the ones on the land. The ones in the city who think they can by not using plastic bags and using paper, for example, really have a negligible effect on any carbon cycle. The ones who think they can can't, and the ones who think they can't can."

Making change is, of course, hard intellectual work. Phil is out to change the stereotype of the "dumb farmer," the uneducated, plodding, backward simpleton so often portrayed in the media. To him, agriculture is joyfully complex, like nature. There's some truth in the dumb farmer stereotype, though: conventional agriculture dumbs down farming to synthetic solutions. Farmers don't have to practice complex thinking to apply pesticides or use GM seeds. The farmer and rancher who can produce healthy, sustainable food in concert with nature, using the sun's wealth, is the farmer of the future in Phil's eyes. That stereotype of the dumb farmer, he says, will go away as more producers turn to regenerative systems. "And I think that guys that can produce real food are going to get handsomely paid for it someday," he says.

Phil sees the country's food producers as the nation's foundation, the stepping-stones on which society is built. Abundant, healthy food encourages a stable society, allowing people to devote their lives to other pursuits besides food production. But if we destroy the land, we have no food and therefore no society, which Phil demonstrates with an example. He mentions two artists who came to the ranch a few months ago to photograph the buffalo. Their art is not necessary for human survival in physical terms, but it helps us be better human beings on so many other levels—and they have the freedom to pursue that kind of job in part because our food supply is stable enough for the time being. "All jobs are important," Phil says. "Like the guys who were out here taking photographs, right, making a living selling photographs to magazines and photo galleries and whatever else. Those are good things. But it all comes back to food and land—and I'm not saying our job is more important—but if we don't do our job right, there isn't room for others."

In other words, food allows humans to flourish to their fullest potential. We know that conventional food production is inherently unsustainable, that someday the system will collapse. If it does, we will have to reconsider how our society functions. Or we can prevent the collapse by creating a new system, one built on the sun's wealth and that values the health of the land over paper dollars.

PART THREE
Organic Regenerative

11

The Surfing Farmer

"When you come back in a month, those flowers will be taking over and you'll *flip out.*"

From under his pale straw hat, Kevin O'Dare (KO, as everyone calls him) grins at me across a raised, plastic-covered plant bed. He's squatting, punching two-inch holes in the plastic with a handheld tool. I'm also squatting, my notebook balanced on my knee. In a few minutes a farm employee will plant flower seedlings in the holes—nasturtiums, bachelor buttons, pansies.

I smile at the phrase "flip out." Flip out isn't typical farmer lingo, and flipping out over edible flowers isn't typical farmer behavior, but I am not on a typical farm. I'm on Osceola Organic Farm in Vero Beach, Florida. When I say "in" Vero Beach, I mean it; the farm is one mile from the Indian River Mall, about four miles from Interstate 95, and roughly seven miles from the Atlantic Intracoastal Waterway that separates the mainland from the barrier island. A housing development leers across the gravel road that is soon to be paved, erasing one of the last vestiges of rural life. Another housing development is slated for construction nearby. Though the farm is quiet—I can hear birds singing in the lemon trees—and the dirt smells wet and rich, I know Starbucks is just down the road.

And then there's Kevin, the most atypical component of this atypical farm. I know farmers, and Kevin is unlike any I've encountered. He wears an army green T-shirt that says "Push Don't Pollute," a phrase I learn later is the brand name of a type of longboard wheel bearing. No John Deere T-shirt for this guy. Kevin also looks, well, *wilder* than any farmer I've met. He sports crooked bottom teeth and curly hair that's orange but giving way to gray. White hairs reach out at me from between his eyebrows, which are bushy and unkempt in an endearing grandfather-ish way, although Kevin is no old man. He's sixty and in excellent shape, thanks to his preferred form of exercise: surfing. He owned and operated a surf shop, Sun Spot, in Vero for twelve years before becoming a farmer. "If there's surf, I'm gone," he says, pointing east toward the ocean.

While running the surf shop, Kevin grew organic food as a hobby at the home he shares with his wife on the barrier island.[1] "I had people come to my yard and say, 'We've never tasted food like this,'" he says. That's probably because it contained a lot of phytonutrients like amino acids, antioxidants, and other defense-related compounds. Organic fruits and vegetables typically contain 10 percent to 50 percent more phytonutrients than conventional produce, which is one reason organic food tastes better, scientists suggest.[2] When farmers use agrochemicals, plants produce fewer phytonutrients because suddenly they don't have to defend against diseases, weeds, and pests. More phytonutrients equal more flavor.

Taste is also related to highly mineralized, biologically rich soil. Delicious food usually originates in this kind of soil because the plants have access to a variety of nutrients. When conventional farmers use synthetic fertilizer and grow in monocultures, though, the soil lacks the minerals and organic content that translate into flavor.[3] It's like the buffalo on Phil's ranch. When they eat grass from nutrient-rich soil, they put on weight and stay healthy, and the meat tastes good, too. When we eat nutritious food, we stay healthy as well. The problem is that our food has declined nutritionally since the advent of conventional farming, a decline that coincides with decreasing soil fertility—and nutrient-poor food is a possible reason for rising obesity levels in America. Per capita, we're consuming far more calories than we did in 1950, and some researchers believe we're eating more because our food lacks nutrition.[4] We're stuffing ourselves in an attempt to meet our dietary needs, all because we've depleted our soils to the point where the food growing in them is nutritionally empty.

When Kevin realized he had produce-growing talents, the next step was scaling up. "I read for two to three years about farming," he says. "I read hundreds of books. I traveled a lot to go look at other farms before I started this farm, so I knew what I wanted to do." Kevin purchased an overgrown grapefruit farm in 1993 and has been growing organic produce on eight acres and organic citrus and avocados on two acres ever since. "We've got these avocados that are off the charts," he says later, handing one to me. They are a special kind I've never seen in any store, a cross between a Hass and a Florida, with skin so thin they couldn't be shipped even if he wanted to do so. Florida avocados are too watery, he says, and the Hass prefers drier conditions, so he started growing a hybrid. "But any avocado is better than none," he says with that grin again.

Kevin managed to survive the start-up period that can kill small farms, that frightful time when more money goes out than in. Ranchers and farmers who convert conventional land to organic like he did face a special challenge: they spend their first three years growing organically but are prohibited from receiving USDA organic certification until the fourth year. This mandatory waiting period detoxifies the land—which is necessary for crops and livestock to be truly organic—but it also means lost revenue. Producers grow organically, but have to sell their wares at conventional prices. That's also when soils are recovering and yields aren't as high as they'll be when soil fertility returns.

Today Osceola Organic Farm includes a greenhouse for starting seedlings, a walk-in cooler, a wash line for cleaning and bagging produce, and a one-room farm stand for selling it. Kevin points out reclaimed or donated items: the buildings are made with old telephone poles and recycled Florida scrub pine, a local chef contributed several used stainless-steel sinks, and the greenhouse is made of old swimming pool enclosures. Long raised beds run north to south, white plastic covering them like crisp bed sheets to create a white-and-black stripe pattern against the dirt. Kevin has been using plastic for about ten years and loves it. He and his workers used to spend hours pulling weeds by hand or sorting them out of the lettuce once it was harvested. The time loss was killing productivity. And there were a lot of weeds: because South Florida doesn't have cold winters that kill them, they grow year-round. A friend suggested trying plastic, and the problem suddenly became manageable since most weeds can't grow through it. "It won't eliminate weeds, but we get a jump on them," he says. "The other thing we didn't realize, a side benefit, is it actually keeps fertilizer right in place." With plastic covering the soil, fertilizer can't leach away with rainfall.

More benefits came to light: because the plants stay cleaner, diseases do not travel as easily from the soil to the leaves, which is important since organic farmers can't use herbicides. In the center of the beds, under the plastic, drip tapes—flat tubes with evenly spaced holes—irrigate the soil. Water doesn't evaporate through the plastic, so Kevin can water less often. Still, plastic is no silver bullet. Kevin pulls a nutsedge plant out of the ground and says this particular weed can poke through the plastic or spread underground. Native Americans, he says, with the plant in the palm of his hand, used to dry and grind the nuts of the nutsedge for flour. Kevin is optimistic about weeds and doesn't think they are necessarily evil, just something he tries to suppress through his management. "Half these weeds that grow out here by themselves are probably the cures to cancer," he says. As he tells me a few months later, "Weeds have been here for hundreds of millions of years. We're actually putting foreign plants into where they are."[5]

But isn't plastic an input, and an oil-based one at that? Is it a truly sustainable practice, and could it break down into the soil? Later I put these questions to Kevin, who tells me the benefits of plastic outweigh the drawback of its oily origin. And there's no fear of plastic breakdown, he says. "Being an organic farm, we have to pull the plastic up. We're not allowed to let it stay on the ground and break down, because it's a petroleum-based product," he says. "But it's recyclable. We've pulled it up, put it in our recycling bins, and the guys take it away. If something is recyclable, I'm all in."

"That's about as perfect as it's going to get," he goes on. "It's not a perfect world. The plastic thing has made my life easier. I don't know if we'd be around if it wasn't for that because our weed labor was 60 percent of our time, and that's way too much in any type of business." To me, Kevin's use of plastic seems like a balanced approach to modern regenerative agriculture: use technology that makes sense for your environment, like plastic does in Vero Beach, but rely on time-tested biological methods, such as crop rotations, in most instances. Plus, the amount of fossil fuel required to manufacture the plastic is far less than what conventional farmers use in the form of synthetic fertilizer, diesel fuel, and agrochemicals, or the fuel Kevin would use if he controlled weeds by cultivating with a tractor instead.

"We try to work with Mother Nature, not against her," he continues, punching another flower hole into the white plastic. Instead of spraying for pests and plant diseases, he uses intercropping, or planting different crops in close proximity. Unlike monoculture farming, intercropping encourages insect and soil life and disrupts disease spread. "We're mixing all that stuff together, like Mother Nature does, where you have companion planting and a break up of different crops to keep the insects at bay," he says. "When you're planting different stuff all around, you're breaking up the monotony of everything and it's hard for diseases to move on and some pests, too, because you're putting a different thing in the way of them that they don't understand or can't get through." He also uses a process called solarization. During the off-season, summertime in Florida, the resting soil reaches temperatures of up to 150 degrees, which "cooks" pests and diseases and reduces their appearance in the growing season.

To replenish soil fertility, Kevin plants cover crops, a measure that also disrupts the weed, insect, and disease cycles. Certain cover crops break up and aerate the farm's heavy soil with their strong, deep taproots. Cover crops also fix nitrogen and add organic matter. The rotation of cover crops and production crops is an organic operation's main way of paying the soil bank, along with adding compost and biologic fertilizers such as manure.[6] These natural nutrients remain in the soil longer than synthetic nutrients would, which leach away and end up in water below and above ground.[7] Knowing this, conventional farmers often apply too much, which means extra cost and more pollution. Kevin's organic methods are a two-way savings: he lets the farm produce its own nutrients, and he keeps those nutrients longer. Again, nature works, not him.

Kevin eschews mechanization for the most part. He keeps a small tractor and a pull-type implement for creating the raised beds, but that's it. His approach to machinery is much like his approach to plastic: use what makes sense and does more good than harm, but ditch what isn't necessary. He and his employees do most farm tasks by hand—planting, weeding, and harvesting, for example. "They have all this fancy equipment that's really expensive," he continues. "By the time I set up something to do this"—he gestures at the raised beds—"we can have this punched and be done." One's feet are the best way to get around the farm. It takes about fifteen minutes to walk the perimeter, a nice stroll past palm and fruit trees. "If I was a little bit bigger, I might get some mechanical wheels or something, but for now we walk," Kevin says. He borrows a neighbor's compost spreader to disperse the 175 tons of new soil he adds each season from his giant compost pile. He would rather borrow than buy equipment that he only uses a few times a year, and he helps his neighbors lay beds in exchange. This borrowing reminds me of Revolution-era times, when agriculture was a community effort, with community benefits. Shared use of plows and other devices was common, as was the shared use of a pasture in the center of a village (from which we have the term "village green"). "We've always been on a budget," Kevin says. "We beg, borrow, never stole. The Bible says, 'Ask and you shall receive' and I put that to the test out here!"

Speaking of stealing, my favorite part of the farm is the sign propped against the farm stand cash register that reads in all caps: WE ARE NOT

HERE NOW. PLEASE HELP YOURSELF AND PLACE MONEY UNDER THE SCALE. REMEMBER GOD IS WATCHING. THANK YOU. This sign is fifteen years old. When Kevin or his employees are not in the stand, which is most of the time, people use the "honor system," leaving cash or checks underneath the scale next to the register and writing how much produce they take on a yellow legal pad. One would think this honor system would invite stealing, but thefts are rare, Kevin says. He calls the scale his "magic scale" because of the money he finds when he tilts it up. "People walk in here at twelve o'clock at night and pick up what they want," he says. "The only time I close those doors is during a hurricane."

12
The Mission

Kevin continues punching flower holes and I continue sweating. It is 8:30 a.m., eighty-two degrees and climbing fast. A bank of blue rain clouds looms to the southwest and thunder murmurs in the background. I hope for rain or at least shade. "We sell the crap out of them," he says proudly about the flowers. Kevin adds that they are not only popular but also profitable, partly because he has to plant them just once a year. In contrast, the farm's main crop, lettuce, is planted weekly during the growing season, between five thousand and eight thousand new plants a week. "We harvest millions of pieces of lettuce a week. I mean millions," he says. The flowers seem easy by comparison, a beautiful addition that happens to pay. "It's a good business, a small niche market," he says about the flowers. "We do a little bit of this, a little bit of that, and it all adds up at the end of the year."

A little bit of this and a little bit of that means: arugula, basil, green beans, beets, broccoli, cabbage, Swiss chard, chia seed, chives, cilantro, dill, eggplant, kale, lemongrass, lettuce, microgreens, onions, oregano, parsley, peppers, rosemary, strawberries, tarragon, tomatoes, turnips, wheatgrass, and anything else Kevin decides to experiment with. He uses "lettuce" as a blanket term for the many varieties of leafy salad greens he grows. There is a lot of plant diversity on Kevin's farm, the kind that used to exist on most American farms but disappeared during the industrial movement. When Kevin first started farming, mentors counseled him not to even attempt lettuce; everyone said Indian River County, situated about two-thirds of the way down Florida's east coast, was too hot for tender lettuce plants. In keeping with his beg and borrow theory, Kevin nonetheless accepted a free bag of organic lettuce seeds from a friend. The stubborn seeds refused to grow in his greenhouse, as his mentors had predicted, so he scattered them over an empty bed just for fun.

"They started growing. Ten different types of lettuce," he says. "One of my wife's friends owned a deli on the beach, and she came out here just to look around one day. She said, 'You're growing spring mix! We can sell the crap out of that.' I said, 'Oh yeah?' I didn't know. So we gave it to them and they loved it. We went to the biggest restaurant in town, Ocean Grille, it's been here since the '30s. I brought it there and they jumped on it right away, and we went from there." The next year, Kevin grew more spring mix (intentionally), cut samples, and dropped them off at eight of the swankiest beach restaurants. "By the time I got back to the farm, there were three or four voicemails on the machine," he says.

Those voicemails signaled a major transition for the farm, from a simple farm stand business to restaurant clientele. Before the restaurants got involved, Kevin sold everything from the stand (Vero Beach didn't have a farmers' market then). The change came just in time: people weren't coming down that dirt road often enough for Kevin to turn much profit, and it was getting to the point that he might have to quit farming. "[The restaurants] saved our butt as far as making money and being able to be fiscally sustainable from year to year," he says. "We took that, and that's what keeps us going." And it all started with a few randomly scattered lettuce seeds.

Strangely, a similar phenomenon occurred with kale. One day Kevin spotted a kale plant growing on a fallow bed. He describes it as being "big and beautiful without any help." How it came to be there, he had no idea. Perhaps a lost kale seed had found its way into a bag of lettuce seeds? Or maybe its origin was something greater? He snapped a picture and posted it on Facebook. "I thought about it after I posted it and another week later I said, 'You know what? That's God telling me I should plant kale there if it grows that big right there.'" And Kevin has grown big, beautiful kale ever since.

Both stories remind me of the biblical parable of the sower, who scattered seeds on four types of ground. Some seeds fell on hard soil and didn't sprout at all, while others landed on stony ground, where they germinated but died shortly thereafter. A few seeds ended up among thorns; they grew a little but the thorns choked them out. The lucky few seeds that fell on rich earth thrived and became fruitful. The biblical lesson is to make one's heart like rich earth, where faith has a good environment for taking root. Kevin did what the sower did—he scattered seeds on rich earth. But the lettuce and kale that grew were not miracles, as tempting as it might be to think so. The farm's fertility is not an accident or some kind of biblical blessing. It's a result of Kevin's hard work, the mental and physical energy he expends daily. Farming organically requires complex thought, careful attention to natural cycles, and a holistic perspective of the world.

In fact, Christian thinking has inadvertently contributed to humankind's abuse of the environment by encouraging the dominion narrative. The dominion narrative rationalizes human authority over the natural world, and it's a philosophy ingrained in most Americans whether they are conscious of it or not. Scholars agree that the dominion narrative arises from the Judeo-Christian tradition, specifically from the book of Genesis in which God instructs Adam to subdue the earth and provide dominion over it. The connection between Genesis and human control over nature was first articulated in Lynn White's seminal 1967 essay "The Historical Roots of Our Ecologic Crisis." White exposed the connection between anthropocentric Christianity and environmental exploitation and degradation, showing that people had interpreted God's command to subdue the earth as permission to do whatever they wanted to the natural world.[1] Human authority over the natural world sanctioned by God—a license for destruction.

The view of humankind as dominant over nature, however, is not associated with the Judeo-Christian philosophy alone. Rather, it is a view deeply entwined in both secular and religious Western thought. Ecocritic Paul Shepard linked the dominion narrative concurrently to Platonism, Christian theology, the mechanization of society in the industrial era, capitalism, and the urban worldview of thinkers like Francis Bacon, Thomas Hobbes, and Karl Marx.[2] So the human-over-nature view has been fairly unavoidable for a long time. The dominion story is especially present in Western life and continues to be our primary cultural narrative, particularly when it comes to managing our agricultural lands.[3] In a world where food originates at the grocery store, not from the work of our hands, and where homes stocked with appliances, computers, and temperature control systems make our lives convenient and comfortable, it's easy to forget what it means to be creatures of the natural world and thus we embrace the urge to control it rather than live with it. We're like indoor cats: wild somewhere in our hearts, but so domesticated that we couldn't survive long outdoors.

If humankind controls nature, as the dominion narrative says, then humans can't also be part of nature. We are outside of it. And if humans don't see themselves as part of the natural world, then its destruction is less troubling. That's one reason conventional farmers typically do not see a problem with their methods; to them, the land is a commodity used to run their business and little more. Their survival depends upon it only in financial terms. Aldo Leopold wrote about the consequences of this view in *A Sand County Almanac*: "Conservation is getting nowhere because it is incompatible with our Abrahamic concept of land. We abuse land because we regard it as a commodity belonging to us. When we see land as a community to which we belong, we may begin to use it with love and respect."[4] Kevin and other regenerative farmers take the "community to which we belong" view, and they understand that their land is part of an ecological whole in which they can't manipulate one thing without impacting everything else. Kevin can't simplify or control nature's diversity, things industrial agriculture seeks to accomplish with its monocultures, fertilizers, and sprays. Instead he teams up with nature as a member of the wider ecological community, and he feels responsible for that community's well-being.

This responsibility is one reason he chooses to grow organically. "I'm so true to the organics thing," he says. "People tell me they are organic and I say, 'Are you certified?' If they say no, then I'm like, 'Well, you're not organic.' They say, 'We do everything organically.' But I can walk through somebody's farm and say, 'That's not organic.' A lot of people *say* it." Kevin has worked as an organic farm inspector since 1995, conducting more than a thousand inspections all over Florida. Farms that claim to be organic but lack certification probably aren't truly organic in his experience. Even on certified farms, he finds occasional rule violations. Good intentions are not always enough. For crops, USDA certified organic means promoting soil fertility through crop rotation, tillage, cover crops, and manure or other natural waste. Pests, diseases, and weeds must be controlled physically, mechanically, or biologically; if those methods fail, farmers can use approved substances. Organic farmers must also use organic seeds (when available), and they cannot use synthetic fertilizers, pesticides, or chemicals (except those on the "safe" list), genetically modified seeds, sewage sludge, or irradiation.[5] Farms undergo periodic inspections and keep immaculate records. "It is a little bit more paperwork, but it's worth it," Kevin says. "We have guidelines that we follow. They are black and white. I sign on the line to follow those guidelines and be certified organic, and the rules are right there. There are no gray areas anymore. It's pretty easy to follow if you want to."

Kevin ate organically years before the National Organic Program was established, but unless he knew the farmer or food processor, he had no way of knowing whether the food was really organic. A national organic standard ensures consistency and purity, he tells me, which is beneficial for buyers. Without a standard, he says, the consumer has no way of knowing what, if any, qualifications a product has met to earn the label. "Having a National Organic Program is a good thing," he says. "When you see that seal, whether it's in Publix or Whole Foods or wherever you shop, you know it's for real." He distrusts the integrity of most organic produce grown outside of the United States because the oversight sometimes isn't as strict. His position on certification is firm: "Everybody says they're organic. Everybody. It's a buzzword. It's hot right now. You aren't going to hear anybody say, 'I'm GMO!' But if they are not certified organic, they are not organic."

When Kevin says this, I think about Phil's stance on certification and how, if they met, they would probably argue about it. Kevin is as stubborn about the benefits of certification as Phil is about the drawbacks. Their differing opinions highlight the larger struggle within the regenerative agriculture world about rules and who sets them. Organic agriculture sounds simple—do not use agrochemicals and GM seeds, do use natural processes that conserve soil and water—until you get down to the brass tacks. What about pesticides derived from natural ingredients, like neem oil from the neem tree? Most farmers agree something like that is acceptable, but some purists don't. Or the case of our organic steer, who can pig out on organic corn in a CAFO but remain certified. Most people consider that practice harmful to the animal and the environment, even though the steer remains technically organic. Others have no problem with organic CAFOs, including the giant corporate food companies that source meat from them.

Kevin advances a moral and health-focused argument in favor of certification. People have many reasons for selecting an organic product, he says, and often those are connected to remedying a disease. If food companies or farmers sell a product that isn't actually organic, but contains "poison" instead (by this he means nonorganic chemicals or ingredients), then the results could be disastrous to the person's health. "If they are recovering from cancer and you sell them something that really isn't organic, maybe it will never affect them—but maybe it could," Kevin says. The deception is also an affront to the consumer's personal liberty and his or her philosophical commitment to supporting organic growers and practices over industrial ones. He believes an important part of the organic grower's responsibility is honoring customer intentions by providing a product that is what it claims to be. "It's the consumer's choice and they are making up their mind which way they want to go, and you have to follow that," he says.

One way inspectors confirm the integrity of organic growers is by scrutinizing their activity logs. Farmers log everything they do, such as what tasks they complete each day and what substances they use, all the way down to the hand soap that workers wash with before handling produce. Every material must be permitted for use in organic production. Products must also be approved by the grower's certification organization. If Kevin wants to make a change as minor as using a different brand of sanitizing solution, for example, he has to inform his certifier ahead of time. Farms are regularly inspected to ensure compliance. Inspections take about four hours at Kevin's farm, two for combing paperwork and another two for visual inspections.

He also keeps precise seed inventories, because the seeds he plants must also be certified organic. Seed companies test their seeds twice a year for accidental GM contamination, a process that makes them extra expensive for growers. Kevin records what seedlings are planted when and where. The result is an audit trail stretching from the seeds to the vegetables, ending at where they are delivered for consumption: an invisible thread linking farmer to consumer. And if something goes wrong—a bag of lettuce is contaminated, for example—the problem is traceable.

It's a lot of hoop jumping, but no more than conventional growers do to enroll in the USDA's farm program and other subsidies. Kevin doesn't seem to mind. In fact, he is excited to share what he does with inspectors and anyone else who might be curious. He welcomes visitors to the farm, inviting them to wander the fields at their leisure and watch the produce being harvested. One can only learn so much from logbooks and long walks, however. A better indicator of the quality of the produce from Osceola Organic Farm literally walks through the door. A woman who lives down the road enters the farm stand, where Kevin and I have retreated for a few moments of shade, and asks when it will reopen for the season. She's moving soon and wants to get a last bit of produce before she goes. She calls the farm "one of the hidden secrets of Vero." About the farm's produce she says, "You just can't beat it. It lasts. Sometimes I have to shop at Publix, but it's bad in two days." She makes a "blah" sound and waves her hand dismissively. Like many other Americans, this woman is fed up with the tasteless, half-decayed produce in the grocery store. Big chains like Whole Foods, Wal-Mart, and Publix stock fruits and vegetables, including organic items, from national distributors, which usually source from megafarms. This system adds fuel miles to the food, and it forces farmers to harvest fruits and vegetables before they're ripe so that they don't spoil, which means they never achieve their full flavor potential. After days on a truck, even the organic produce loses its freshness. Local farms like Kevin's are much better at feeding their communities and regions without the fuel use or produce breakdown. That's another reason locally grown food is usually tastier and lasts longer.

Kevin has nothing to offer the woman; he's still a month away from the first harvest of the season. She bids him good-bye and says she'll miss him and the vegetables.

13
The Plants

Every spring on the farm, a magical day comes when the first tinge of green appears in the field, when tender shoots poke through the soil, reaching for the sun. "Wheat's up!" my dad would say when we'd ride the ATV to scout fields. The shoots fluttered in the breeze, fragile and skinny as young grass. "Corn's up!" he'd say a few weeks later. The corn was sturdier, thicker. He was always supremely pleased to see the little plants clinging wholeheartedly to the dirt as the South Dakota wind blew, underdogs in the fight to survive. When you think about it, the act of seeds sprouting into plants is mind-blowing. I still can't believe it sometimes: a pebble of a seed takes in water (a process called imbibition), which hydrates enzymes and food stores and prompts the seed to produce energy. A root called a radical (I love that term) breaks through the seed coat. The shoot follows, heading toward the sun. When the shoot feels the first ray, it starts producing leaves and turning green. The whole process is called photomorphogenesis.

They look so innocent, those pale green sprouts on America's farms. But not all is as it seems; 92 percent and 94 percent of the corn and soybeans, respectively, in our country's fields are genetically modified.[1] They are human-made creations disguised as nature's corn and soybeans. They contain the genes of bacteria, animals, and plants placed within their DNA by scientists in laboratories. Knowing this, the spring days I spent working for the newspaper, when I admired the fledgling sprouts from the car on my way to an assignment, take on a sinister quality. I, like many others, did not know what alien plant life lurked in the fields.

How did we get here, anyway? In the late 1990s, the plastics-company-turned-agrochemical/biotech-corporation Monsanto released a new invention: seeds that its scientists had genetically modified to survive the application of Roundup, the company's signature herbicide.[2] Roundup, or glyphosate, was at the time the most effective weed and grass killer available to farmers, but, like other herbicides, spraying it would kill the crop along with the weeds. Farmers mainly used Roundup to "burn down" bare fields, or kill weeds before and after planting. When Monsanto created plants capable of surviving Roundup ("Roundup Ready" as they were dubbed), farmers were astounded. Spraying herbicide directly onto corn (1998) and soybeans (1996) achieved far better weed control. The company eventually created Roundup Ready cotton, canola, sugar beets, and alfalfa. Its patent on glyphosate expired in 2000, but it still markets its Roundup Ready seeds. Monsanto has already developed glyphosate-tolerant wheat, which will likely appear on the market in the coming years if federal regulators approve it. GM corn and soybeans became America's most-grown crops, and they still are. More than 50 percent of U.S. cropland, or 163.5 million acres, is devoted to corn and soybeans.[3] Farmers embraced Roundup Ready technology as fast as every other tool they'd been offered by the agribusiness industry. This one, though, proved dangerous in new ways.

Since GM crops could withstand glyphosate only, farmers used that same herbicide at least once a year, but usually two or three times a year. Using so much glyphosate put evolutionary pressure on weeds, and at least ten "superweeds" became resistant to the chemical.[4] Farmers responded by increasing the amount they use, which caused further resistance. Higher application rates and more frequent use also means glyphosate covers the grain we eat in higher concentrations than ever before. Glyphosate residue lingers in our food, especially processed items derived from GM corn and soy.[5] It also coats the feed our livestock eat, and since chemicals become more concentrated as they travel up the food chain, our meat is laced with glyphosate, too. We also encounter glyphosate in the air, surface water, and rain.

Glyphosate is even in the fuel we put in our cars thanks to the ethanol industry. If the 1990s was the era of genetic modification, then the first decade of the new millennium was the era of ethanol. The U.S. government sponsored the development of a massive ethanol industry to suck up the surplus corn caused by increased production. Ethanol plants sprang up across the Midwest in states like Iowa, eastern South Dakota, Minnesota, and Illinois. Demand for corn skyrocketed as more ethanol plants opened, causing prices to double in 2007.[6] Corn farmers cashed in on record profits, which they sunk into expansion efforts to produce even more corn. Ostensibly the project was intended to create sustainable biofuel—sounds pretty good, right?—but the results have been ecologically and socially destructive. Land disappeared from conservation programs. Farmers began growing "corn on corn," or planting corn on the same fields year after year, which increased their need for fertilizer, depleted soil, and further contributed to insect and weed resistance. Ethanol was the final nail in the coffin for farm diversification, as the lure of high profits swayed farmers to convert pastures, barnyards, and hayfields to cornfields. Edge tillage, a practice not used since before the Dust Bowl, returned almost overnight.[7] Edge tillage, which means removing borders between fields, is another way of saying one is farming fencerow to fencerow.

My many miles on the back roads of Iowa, Minnesota, eastern South Dakota, and northeast Nebraska on *Tri-State Neighbor* assignments confirm that edge tillage is back big time. What I saw was "corn creep." Farmers had moved their grain bins and machine sheds in tight clusters around their houses, reducing the farmyard to make way for more corn. Fields had crept closer to the farmhouses, riverbanks, and roads, closing in on the people. I even saw front yards planted to corn. The scenes now remind me of a line from Pollan's *The Omnivore's Dilemma*: "Corn had pushed the animals and their feed crops off the land, and steadily expanded into their paddocks and pastures and fields. Now it proceeded to out the people."[8]

Corn production continues to increase, pushed on by the ethanol industry. Less than 5 percent of the U.S. corn crop went to ethanol production in 2000; the rest went to food and livestock feed. By 2013, 40 percent of U.S. corn went to ethanol, 45 percent to livestock feed, and 15 percent to food and beverages.[9] The ethanol boom meant a windfall for some farmers, but it also drove even more extreme specialization that hurts them when prices fall, as they did in 2014. Reports from the United Nations Intergovernmental Panel on Climate Change, the International Institute for Sustainable Development, the Environmental Working Group, and many others have proven that ethanol is inherently unsustainable. I doubt it will survive as an industry, especially as wind and solar power take hold. When ethanol finally dies, those farmers who invested heavily in corn—those who are no longer diverse—will suffer.

Not long after Roundup Ready seeds appeared, Monsanto invented yet another GM corn variety, this one engineered to produce insecticidal toxins derived from the bacterium *Bacillus thuringiensis* (Bt). A natural, soil-borne bacterium, Bt produces crystal-like proteins (Cry proteins) that poison certain insects that eat it. Hence Bt corn, as it is known, is actually a registered pesticide, as it can indeed kill certain pests, such as the European corn borer and the western corn rootworm. Bt corn became a commonly grown crop, as did Bt cotton—but as with Roundup Ready seeds, overuse caused problems. In the United States, the western corn rootworm is resistant to Bt corn, and many scientists expect other insects to develop resistance in the coming years.[10]

Today corn, soybeans, and cotton are genetically engineered to resist all kinds of other herbicides (like dicamba and 2,4-D), tolerate insects, withstand drought, and have increased yields.[11] Farmers are also growing GM potatoes, apples, sugar beets, alfalfa, canola, papaya, and squash. The industry has always insisted that GM seeds are safe. Whether that's true or not hasn't been thoroughly tested. Scientists who question the safety of

GM food are harassed by agribusiness corporations and often find themselves blacklisted in the academic community. French scientist Gilles-Eric Seralini, a professor at the University of Caen in France, discovered in a two-year study that rats fed GM corn and Roundup developed liver and kidney disease and mammary tumors. When his work appeared in 2012, Monsanto and scientists linked to the company lambasted the study so vigorously that the journal retracted it. Seralini republished his findings in 2014.[12] Other scientists who've reported negative connections between GM food, glyphosate, and human health have been similarly silenced.

Around 80 percent of processed foods contain GM ingredients, and Americans eat *a lot* of processed food. Yet we aren't allowed to label GM products or even examine the safety of something most of us consume every day. This frightens me not only because of the potential human health and environmental consequences, but also because of the extreme control corporations like Monsanto have over scientists, politicians, professors, journalists, all of us. The agribusiness companies that controlled the news at *Tri-State Neighbor* are still trying to do so there and everywhere. We're talking about the silencing of free speech in a country that claims to value it.

In Kevin's greenhouse, however, every plant is what it claims to be, inside and out. Sunlight sifts through the mesh enclosure onto black flats of seedlings on tables. Kevin points at the just-sprouted plants and names them: basil, purple basil, dill, red Russian kale, chard, turnips, beets, lettuce. As is the case with many small-scale organic growers, Kevin's decision to farm organically is rooted in his ecological, spiritual, and philosophical beliefs, and those beliefs include a deep skepticism of GM crops and the processed foods made from them. "I look at GM foods as Frankenstein food because we don't know what it is," Kevin says. "How can you make better food than God? When people ask me, 'What is organic?' I go, 'Organic is the way God intended food to be grown.' That's it. Period."

Kevin is not against science. He is not a religious fanatic, and he does not advocate a return to farming with primitive tools. "The way God intended food to be grown" is his way of describing the philosophy of mimicking nature, not subduing it. Whether they call it God or Mother Nature or something else, many people who believe in regenerative agriculture agree that there is something deeply wrong, something deeply out of sync with the natural order of life on earth, when farmers replace natural processes with chemical solutions and when genetically altered food becomes part of our daily diet. Such actions threaten not only our health and planet, but also the survival of humankind in the future. Kevin also says he wants to avoid degrading resources that the next generation depends on. "I base everything on the future generations. No matter what, we have to do it the right way. If you have kids or plan on having kids, you have to look at it that way," he says.

But can regenerative, organic agriculture provide enough food for a growing population? The most common objection from conventional farming advocates is that organic agriculture cannot "feed the world," something American farmers now feel obligated to do, never mind that they cannot even feed their own communities or themselves. The threat of hunger is one reason we industrialized the food system in earnest after World War II. Agribusiness proponents framed industrialization as the only way to grow enough food for everyone, and they argue that organic agriculture isn't productive enough. The evidence, however, suggests otherwise. In the longest-running organic versus conventional field test in the United States conducted by the Rodale Institute, organic crops consistently show comparable or higher yields than conventional crops.[13]

This side-by-side evaluation, called the Farming Systems Trial (FST), was started in 1981 and continues today, and it is a reliable source of evidence in assessing the long-term potential of organic. Taking a long view like Rodale does is necessary for comparing organic and conventional systems. Other researchers conduct single- or multiple-year field tests, and these typically find that organic yields are lower than conventional yields. Why? The answer lies in the soil. Researchers often plant trial fields on land formerly used for conventional growing. Even if the land was used for other purposes, it's safe to say most wasn't managed organically and isn't exceptionally rich, so test plants start off in depleted soil (just a medium to hold the plant up, as Phil put it) that doesn't hold water and/or contain high enough levels of nitrogen, organic matter, and carbon. The conventional plants receive synthetic fertilizer that compensates for the poor soil, while the organic plants have to wait until cover crops replenish nutrients (a process which takes many years) or until organic manure or compost is applied. The conventional plants also receive chemicals to control pests and insects; resistance in an organic system, however, is built over time.

It's no wonder conventional yields are higher in the first few years of side-by-side comparisons—but then a shift starts to occur. As soil health and pest resistance improve in the organic system, yields increase. The true potential of organic is unlocked with time, and the evidence is undeniable that organic agriculture is not only more productive, but also more resource efficient and profitable. Here are some highlights from the Rodale Institute's thirty-year report on the experiment:

"Over the 30 years of the trial, organic corn and soybean yields were equivalent to conventional yields in the tilled systems."

"Wheat yields were the same for organic and conventional systems."

"Organic corn yields were 31 percent higher than conventional in years of drought."

"Corn and soybean crops in the organic systems tolerated much higher levels of weed competition than their conventional counterparts, while producing equivalent yields."

"Organic farming uses 45 percent less energy and is more efficient."

"Conventional systems produce 40 percent more greenhouse gases."

"Soil health in the organic systems has increased over time while the conventional systems remain essentially unchanged."

"Organic fields increased groundwater recharge and reduced runoff."

"The organic systems were nearly three times more profitable than the conventional systems."

"Even without a price premium, the organic systems are competitive with the conventional systems."

"After thirty years of a rigorous side-by-side comparison, the Rodale Institute confidently concludes organic methods are improving the quality of our food, improving the health of our soils and water, and improving our nation's rural areas. Organic agriculture is creating more jobs, providing a livable income for farmers, and restoring America's confidence in our farming community and food system."[14]

14
The Lifestyle

Half of the edible flowers Kevin grows are not even eaten, but used for plate decoration at fifteen of Vero Beach's most exclusive restaurants, ten of which are located in private country and residential clubs. One account is John's Island Club, which in 2016 was rated the sixth-best country club in the United States.[1] These fifteen accounts purchase about 80 percent of Kevin's produce; he sells another 15 percent at the Vero Beach Farmers' Market and the remaining 5 percent from the farm stand.

The high-end chefs who buy Kevin's produce are paid to create art as much as they are to concoct Michelin-rated cuisine, and edible flowers are only the beginning. Chefs have asked Kevin to bring in bags of orange blossoms to garnish citrus-infused dishes. Some ask that the lettuce be snipped to fit specific salad plates. Many want the spring mix cut, washed, and delivered that morning in time for lunch service. Others have asked that he grow certain vegetable varieties, a request Kevin is happy to accommodate. His operation is small and flexible enough to respond to demand quickly, unlike huge farms such as Ryan Roth's. At two restaurants, Kevin even checks the coolers daily and replenishes stock as needed.

In Kevin's world, there are no salespeople, packinghouses, or distributors. Just four employees work part time alongside him. Because of this, Kevin is one man wearing many hats. He transcribes each chef's order from voicemail, email, fax, or text (he says he'd accept them via Facebook if chefs wanted).[2] Then he helps harvest and box the produce and loads it in his Honda Element, which is more like an office on wheels. He does every delivery in person. Depending on traffic and how long the chefs chat him up on the route, he's done in a couple of hours. "Then I go home and make myself a huge smoothie," he tells me with a grin, and it's back to the farm for cleanup and prep for the next day. When he's not harvesting and delivering, he's busy with the countless other tasks that make the farm function. He works sixty to seventy hours a week during the growing season, but he seems way too driven, or maybe too fussy, to transfer significant tasks to anyone else. He has tried hiring farm managers, but they don't quite understand regenerative farming like he does. They are also not as serious about making sure things are perfect—and this is something Kevin is *very* serious about. At one point during the morning, he spots a tray of unplanted veggie seedlings sitting out in the sun, left there for a few moments by an employee. Kevin shakes his head and says he would never leave a tray out, that he'll have to talk to the person about that.

I ask Kevin to be honest about something: does the farm overtake his life? Does he feel like he never had enough time with his son, who is in graduate school now, or his wife? I ask because my father has always struggled to balance time and money spent on the family and the ranch. The ranch wins in most cases. Before multiple sclerosis stole my mother's ability to walk, she talked of visiting the Grand Canyon, the Smoky Mountains, and even Europe. My father did take her camping at Yellowstone in '80s, and back in the '90s it rained so much that she convinced him to take a trip to Iowa to visit her sister (and I suspect my father only agreed because of all the corn and soybeans he could observe on the drive). Those were his only extended stays away from the ranch until 2015, when he came to Florida for my graduate school graduation. At sixty, he still works seven days a week. Aside from the occasional long weekend, it's mostly work and little play.

This is what conventional agriculture does to families. Motivated by the constant fear of being driven out of farming by bigger operators, my father runs on John Ikerd's treadmill of industrialization like a hamster on a wheel. The sad thing is that the running is only exhausting him. The sophisticated equipment, livestock antibiotics, synthetic hormones, agrochemicals, and GM seeds are generating more cash, but also much higher expenses. Because every other conventional farmer has adopted these practices and therefore produced more, prices remain low—again demonstrating Eric Schlosser's fallacy of composition: the misguided belief that one individual's advantageous practices will still be beneficial if everyone else decides to do the same thing. Because herbicides have created chemical-resistant life forms, he must spend more time in the sprayer every year. And because he expanded his wheat and corn acreage, he had to make huge investments in seeding, spraying, and fertilizing equipment—and he still doesn't have all of the machinery necessary for row crops at this scale, so he's forced to hire expensive agribusiness contractors.

I asked him once about the fact that he can't harvest the wheat, his favorite crop. Harvest is the culmination of a season's work, a moment of triumph in a year of endless labor. First my father said it's nice to have someone else do it, that it's quicker and safer with their massive harvesting machines. The wheat is in the grain bins before a hailstorm or grasshoppers can get it. But then again, there's a joy in "driving down the field and seeing what's growing and seeing what you need to do," he said. "Someone else gets the joy."

"I miss doing the things that I used to do," he continued. "You become more of a manager than an actual worker. I guess it's what you wanted, but once you get there it's like . . ." His voice trails off and he falls silent. *More of a manager than an actual worker*. His admission reminds me of Ryan Roth. I can see where our farm is headed if nothing is done to stop the relentless expansion. My father and I don't speak for several moments, both of us holding our respective phones to our ears thousands of miles apart, and I wonder if he is remembering how the wheat kernels used to feel as they poured through his fingers on harvest day, back when the farm was smaller, the warm breadlike smell left behind on his hands.

Before me stands Kevin, who has farmed the same ten acres for twenty years. Clearly he did not drink Butz's "get big or get out" Kool-Aid. He considers my question for a moment, then says, "You have to have a balance in life. Stephanie, when you have kids, hold your kids tight, hold them as long as you can, because they grow up real quick. I thank God there was a balance out here where I got all my work done, but I was also able to be with my kid and my family a lot." He tells me how the farm allowed him to watch his son's football practices and attend games. He would take a few hours off, then finish his work in the evening. Kevin encourages his employees to follow his lead and prioritize family over work, even if that's not a favorable business policy. Kevin can't afford to offer employees paid leave (a benefit industrial-sized farms rarely offer either), but he can offer flexibility, understanding, and the promise that their job will be there when they return. "I told them no matter what, if there are family problems, go," he says. "I've probably lost money because of that, but my value, as far as I'm concerned, is that we do everything for our family, or we should be."

A lot of farmers, whether on large-scale farms or smaller ones, conventional or regenerative, work for the sake of their families like this. Ryan Roth does. My father does. Maybe both men see their large farms as security for their children. Maybe they see their long hours as sacrifices. Yet the way Kevin farms allows him to have the best of both worlds, a successful business that meets his family's needs and allows time for a life outside of work. I can't help but see Kevin's approach as working smarter, not harder. He seems . . . happy. Happy in a way farmers and ranchers aren't in places like South Dakota today. Conversations in the Bison feed store, where I accompany my dad sometimes when I visit home, are more like laments about all the work that isn't done. The tractor is broke down in the field, they grumble. The calves aren't branded because they're behind on planting wheat. Can't spray because of the wind. There's a general feeling that wheels are spinning but no one is making much progress.

Crop prices are another common complaint. With no way to differentiate their crops or livestock—everything is standardized in conventional agriculture—producers have no grounds for demanding higher prices. I think of Ryan Roth: *Cabbage is market price. You have no control over that whatsoever*. Not in Kevin's model, though. Marketing products directly to consumers and chefs not only allows him to set the price and thus cover his expenses, but also to provide a fresher, tastier product, which keeps buyers coming back. "We don't do wholesale," he says. "That would kill me. We sell our lettuce for ten dollars a pound retail, but wholesale it's eight dollars a pound. I have people come up to me and say, 'We'll buy it all day long for five dollars a pound.' I say, 'No, you won't. Not here.' They say, 'We can get it from California.' I say, 'Yeah, but that stuff is six or seven days old.' There's a big difference in quality. There really, really is." I couldn't agree more: when I participated in a local community-supported agriculture (CSA) program a few years ago, receiving a share of vegetables every two weeks, I saw firsthand the difference between fresh produce cut that morning and produce shipped from places like California, Argentina, and Mexico. Both were organic, but the local greens and vegetables stayed crisp two weeks or longer and tasted way more delicious.

Another benefit of selling directly to customers: Kevin knows exactly where the fruits of his labor go, and I imagine this is satisfying knowledge. I think about my dad, who sells his crop of calves at an auction barn every year. The highest bidder takes them away to a CAFO. My father will never know the man, woman, or child who eats the meat, whether they like it, whether they appreciate it. He works year in and year out for a faceless, nameless consumer, never feeling the satisfaction of positive feedback. Kevin, by contrast, hears thank-yous from chefs and customers. He sees the smiles on their faces, listens as they describe the delicious greens, hears the gratitude in their voices, takes note of their suggestions for improvement. The story is complete. Kevin experiences fulfillment, while conventional producers experience what seems to me like an emotionless transaction.

The connection between consumer and producer is about much more than fulfillment, though. It's about understanding. When people left the farm, they severed their connection to food production. They became consumers, not producers. In the city, food comes from the grocery store, not the garden, corral, or chicken coop of the local farmer, which eliminates the need for dialogue between the two. This divorce between farmer and consumer remains today. When consumer and farmer become estranged, Wendell Berry writes, "The consumer withdraws from the problems of food production, hence becomes ignorant of them and often scornful of them; the producer no longer sees himself as intermediary between people and land—the people's representative on the land—and becomes interested only in production. The consumer eats worse, and the producer farms worse."[3] When farmer and consumer separate, both suffer, and so does the land and the quality of our food. Restoring that connection is an important element of regenerative agriculture, and for the emotional and psychological well-being of our nation's farmers and ranchers.

15

The Consumer

Given the entrenchment of conventional agriculture, is there any hope for the next generation? I ask Kevin. "There is hope, but we have to change things," he says. "I do believe there is always hope. We have more people interested in our food than twenty years ago."

We can see a revived interest in food all around us—witness the organic movement that started in the 1960s and grows stronger every year, the celebrity chefs on Food Network touting the benefits of fresh and local food, the popularity of farm-to-table eating. Consumers care about the origins of their food much more than they used to. But is consumer interest and awareness enough? Can the consumer actually create lasting change in the food market, the kind of change that will result in the widespread adoption of regenerative farming?

Ecological modernization theory argues yes, that consumers can inspire steady movement toward ecological sustainability within a capitalist democracy. How? "Environmentally minded free-market entrepreneurs acting in concert with the state and spurred on by consumers and movement organizations would yield a system that favored the development of technologies and social practices that would protect the environment, while still allowing for growing prosperity within a largely capitalist framework," writes Dr. Brian Obach of the State University of New York at New Paltz.[1] In this theory, the government steers private actors toward sustainability through flexible and locally appropriate measures rather than via a "command and control" approach over the entire market. Characteristics of the market economy, such as competition, can potentially be compatible with a move toward environmentally sound practices, Obach claims. Finally, social movements can generate ideas, policies, and support rather than just critiques from outside the system. Ecological modernization theory is a model in which environmental sustainability can coexist with marketplace profit, a view championed by writers such as Thomas Friedman and Al Gore.

When Obach applied ecological modernization theory to organic farming in the United States, he found that organic's development is in many ways consistent with that theory. People within the organic movement are usually constructive participants rather than outside critics. The promise of profits spurred entrepreneurs to create organic-oriented businesses, and "while movements played the role of educating the public and spurring innovation, it was the market and savvy, socially conscious entrepreneurs who allowed organic practices to become widespread."[2] The state's role has been more complex. Obach argues that the implementation of national organic standards is not command and control because the government worked with growers, movement activists, and other stakeholders when creating the standards.

But there's another way to look at the development of organic: through the lens of the treadmill of production theory, which says that

> the competitive quest for profit and the corresponding economic expansion that characterizes capitalism are not consistent with the earth's finite resources, relatively stable ecological systems, and basic laws of thermodynamics. The central social actors within capitalist democracies—capital, labor, and the state—are all oriented toward economic growth, thus, no significant checks exist to redirect production toward environmentally sustainable practices, even as environmental reforms advanced by a weak or co-opted environmental movement may, on occasion, temporarily slow the treadmill of production.[3]

Treadmill theory is inherently pessimistic about a capitalist society's concern for anything but profit. Social or environmental issues will never outweigh profits unless the power of law or extreme consumer backlash force companies to change their practices. All consumers can hope for under this theory is a temporary slowing of the treadmill. As Obach writes, "From this perspective organic agriculture is, at best, an authentic social change movement that was co-opted by the dominant treadmill forces, who redirected it in order to increase profits and expand production."[4] We know that big industrial farms and food companies did not experience a change of heart when they went organic; their motive was profit-driven, and they continually fight to relax organic rules to increase profits. Obach gloomily concludes that the development of organic fits both models and there is no "winner" in terms of how we can interpret what went on in the past or what might happen in the future.

If macroeconomics are inconclusive, then what about the microeconomics of individual wallets? We have all heard that in America consumers can vote with their dollar. Can't consumers rebel and refuse to buy conventional and GM foods? Yes—and no. As Robert Albritton and others have pointed out, the individual economies of wallets are not very effective tools for change. He writes that consumers are not likely to modify their buying behavior given their limited resources, knowledge, and alternative options. The American recession of 2007–9, the effects of which are still lingering in the form of lost homes and stagnant wages, only weakened people's buying power. Plus, Albritton says, it's unfair to expect individuals to shoulder the whole burden of change: "If the individual is the main focus for change, then change is not likely to be effective, not only because of the unrealistic burden on individuals, but also because we are asking individuals to struggle against the root causes of the problem which remain intact."[5] Of course individuals can and should be involved in changing the food system, but we can't expect them alone to topple a system so entrenched, so well-funded and organized, and so extraordinarily profitable for big companies.

The optimist in me wants to believe in the hope Kevin feels and in the tenets of ecological modernization theory. As a consumer, I feel powerful strolling through the Whole Foods aisle or shopping at my local farmers' market. I can fill my basket with organic vegetables and grass-fed beef, GM-free tortilla chips and locally grown strawberries. But the treadmill keeps rolling in the background, deflating the power I momentarily feel. Profit is the only thing that matters in capitalism, and big companies and farms will not change until it becomes profitable for them to do so. We've seen this profit mentality in other cases, not just in agriculture. For example, the certainty of total environmental destruction and the demise of humankind hasn't significantly reduced fossil-fuel use (see Friedman's *Hot, Flat, and Crowded*). It seems rather naïve to think changing agriculture will be different. As a 2013 United Nations Environment Programme (UNEP) Emissions Gap Report notes, "Bringing about change in agricultural management practices [for] climate or other reasons is not easy. More often than not there are important market- or tenure-related barriers that need to be overcome."[6] Appealing to intangible ideas like consumer health, climate change, or soil loss will not persuade most farmers, because many think regenerative agriculture is a bigger threat to their survival than any of those things.

Money aside, farmers often operate in tight-knit, small communities, where new philosophies are viewed with skepticism.[7] People talk—and what they say can be vicious when someone's actions appear to challenge the status quo. Many farmers fear being shunned by neighbors if they adopt regenerative practices. It sounds old-fashioned to those of us in the city, where few people know or care what we do, but this fear is real in rural America. Phil, for example, has endured years of criticism to his face and behind his back. Some people view a neighbor's move away from conventional practices as a personal attack on their own farms and the community as a whole. If you reject our way of farming, the thinking goes, then you reject us.

But it doesn't have to be this way. Money could play a different role in farming. If farmers were incentivized, change would occur much faster, as we saw with the rapid conversion of farmland acres to corn production over the last twenty years, in part a result of government subsidies. We saw it also with the Conservation Reserve Program (CRP), signed into law by President Reagan in 1985 and still active today. The program offered farmers the option of taking their land out of production in exchange for payments. Farmers switched to conventional agriculture in the first place because of the potential for higher profits. If money was enough to convert them to conventional, it stands to reason that money could also ease the transition back to regenerative agriculture.

As a dangling carrot, money is tempting, but it probably isn't enough. The UNEP report advises that a combination of changes in farming practices, government policy, technology, and market conditions is more successful—taking an integrated approach rather than trying to overcome barriers individually. The most successful policies are attuned to local conditions, they write.[8] The notion of local conditions is a key one to my mind, because it implies learning to work with local environments instead of fighting them. It also implies taking into consideration the context of farming communities and individual farmers when developing strategies for change. Local contexts might sound incongruous with government initiatives that tend to be "one size fits all." In the U.S., though, there is a way to reach communities with county-specific information and assistance: the Cooperative Extension Service (CES).

The CES was founded in 1914 to turn research into practical farm tools and to educate farmers about new or existing farm technology. The CES began serving the nonfarm public as well, most memorably by assisting with seed, fertilizer, and tools for people participating in the Victory Garden program during World War II. The Victory Garden program was so successful that participants grew 40 percent of the vegetables consumed in the United States in 1943. Today, the CES works in six broad categories: 4-H youth development, agriculture, leadership development, natural resources, family and consumer sciences, and community and economic development. Each category includes a plethora of programs for children, teens, and adults, all with the intention of helping and educating communities.[9]

The CES is currently an ally with agribusiness, not with the regenerative farming movement. Land-grant colleges, themselves funded by agribusiness companies, oversee the CES and determine what information reaches the public, much of it from the agribusiness world. As a result, the CES is complicit in the "get big or get out" mandate. Berry notes that CES training was originally intended to assist farms that were, in the government's view, "either too small or too unproductive or both."[10] Berry argues that the notion that small or not-so-profitable farms need to be "fixed" by government programs like the CES further contributed to the "get big or get out" theory—and he's right.[11] When the government pointed to certain farms and said, "You are not big enough and don't generate enough profit—you need government help," the government inherently placed a higher value on big farms.

But voters can change the mission and purpose of the CES. There are roughly 2,900 county and regional CES offices nationwide. We have 3,007 counties in the United States, so most areas are covered already or could easily become so. Redirecting this expansive network so that it provides sustainability training and food education, for example, means we could use an existing system, with existing employees, offices, and resources, to reach almost every county with the message of regenerative farming and thinking. CES programs could help today's farmers and ranchers start the transition away from industrial agriculture, much like the CES once helped households across the country plant Victory Gardens.

Land-grant colleges also offer hope for the regenerative agriculture movement because they have the power to influence the next generation of farmers. But extracting these universities from the grip of agribusiness will be difficult. Many of the nation's well-known schools are land-grant colleges: the University of California, Purdue, the Massachusetts Institute of Technology, Cornell, Ohio State, Penn State and Texas A&M, to name a few of the 106 in existence. The federal government launched the land-grant system in 1862 with the goal of making agriculture a science, educating young farmers and urban citizens, and conducting agricultural research.[12] As federal funding decreased, agribusiness companies saw an opportunity to influence education and research to their benefit. These companies built fancy research facilities (like the Cargill Plant Genomics Building at the University of Minnesota), endowed university positions (like the Monsanto–endowed chair for the Agricultural Communications Program at the University of Illinois), and funded research (like the University of California's Nutrition Department research into the benefits of eating chocolate, funded by Mars, Inc.). They also infiltrated university leadership. For example, David Chicoine, president of South Dakota State University from 2007 to 2016, and current professor of economics there, has also served on Monsanto's board of directors since 2009.

The most troubling issue is how agribusiness influences the outcome of university research, known as the "funder effect." More than 15 percent of university scientists report changing their results or methodologies to please a donor. If researchers come up with unfavorable findings about agribusiness products, then the companies simply block the scientists from publishing their results under patent protection laws. An example from Food & Water Watch's report *Public Research, Private Gain*:

> When an Ohio State University professor produced research that questioned the biological safety of biotech sunflowers, Dow AgroSciences and Pioneer Hi-Bred blocked her research privileges to their seeds, barring her from conducting additional research. Similarly, when other Pioneer Hi-Bred–funded professors found a new GE corn variety to be deadly to beneficial beetles, the company barred the scientists from publishing their findings. Pioneer Hi-Bred subsequently hired new scientists who produced the necessary results to secure regulatory approval.

Evidence of corporate control over land-grant colleges goes on and on.[13] The situation is not beyond repair, though. Land-grant colleges still rely on public funding and are therefore subject to public control. Food & Water Watch (a Washington-based NGO focusing on government and corporate accountability related to food and water) recommends specific actions that would help rid the land-grant colleges of corporate influence: (1) the farm bill should prioritize and fund research that benefits the public interest, (2) land-grant universities should be more transparent about their funding sources, (3) the universities should not allow the lure of profits to determine their research agendas, and (4) academic journals should enforce rigorous conflict-of-interest standards. The colleges themselves also need to enforce conflict-of-interest rules. Professors, board members, presidents—anyone who has the power to influence research and classroom curriculum—should not also work for agribusiness companies.[14]

All this is possible—a radical change in how America supports its farmers through subsidies, a retooling of the CES, a major redirection of the land-grant universities—but the biggest hurdle will be opposing the Big Ag lobbies in Congress. As Kevin says about the behavior of Big Ag, "Once you get on top of your business, you start rolling boulders down the hill so nobody else can climb up." Companies at the top, like DuPont, Monsanto, and Con-Agra, have been rolling boulders at organic farmers for decades in the form of antiorganic media campaigns, and food companies like Coca-Cola and General Mills do this by fighting GM labeling and tougher organic standards.

Even the terminology—organic versus conventional—is part of a boulder-rolling attempt. "Conventional" means socially accepted, usual or established, and based on consent. In a way, this is an accurate word to describe industrial farming: it has become socially accepted and quite established. But "conventional" carries the connotation of normalcy and correctness, and that is exactly why agribusiness disguised industrial agriculture with that particular name. It sounds good to people who don't understand what conventional agriculture actually means. "Conventional" sounds right, like the way things are supposed to be. The norm. In reality, conventional agriculture is a massive departure from farming as we have known it for some ten thousand years. There is nothing normal or long-standing about it.

16

The Farmer Goes to the Table

I'm riding shotgun in Kevin's Honda Element, cruising down the road with about $900 worth of produce behind my seat. We're taking tomatoes, flowers, strawberries, microgreens, beets, parsley, and lettuce upon lettuce to Kevin's restaurant clients. It's quite chaotic—receipts and papers fly about, an empty smoothie container rolls on the floorboards, and boxes tip over and slide around in back. The Element, being box-shaped itself, is the perfect delivery truck/office/command center: roomy enough to haul a day's worth of deliveries, but not too big to be unwieldy on the road. Also, it's Valentine's Day, the busiest day of the year in the restaurant business. No pressure to get these deliveries done or anything.

We head east on a paved road, toward the beach. As we drive, Kevin tells me about the chefs we'll visit. He's known many of them for years; he can tell me what states they've worked and lived in, about one chef's wife whose cancer was cured by eating organic food, and who he used to be surfing buddies with or still is. Some are lifelong friends. The commitment between farmer and chef, then, is there. He has to look each chef in the eye—and by extension, all of their customers—and vouch for the produce he delivers. "We could be best friends, better than we are now, but if I didn't have the quality I wouldn't be here," he says. "Knowing these people, it might have got me into these places, but it doesn't keep me in them. My produce does."

We make several stops on the mainland before crossing a bridge over the Atlantic Intracoastal Waterway to the barrier island. At each stop, I help Kevin unload produce. I carry bags of flowers and boxes of lettuce to the back doors of restaurants too fancy for someone like me to eat in. The act of delivering food to a restaurant directly from a farm is surreal—what I have in my hands was harvested just a few hours ago, and diners will eat it for lunch or dinner today. Not so at most restaurants, where food is purchased from a wholesale supplier like Sysco or Cheney Brothers. "They can call one company and get everything they want, from first aid kits to fire extinguishers to hand cleanser to green beans to steak to juice. Everything. Organic, too—anything you want, all with one phone call," Kevin says. Sourcing from a local grower like him requires chefs to work harder—he is an extra person to deal with for chefs who already have a lot on their plate, so to speak.

It's useless for chefs to source food locally or include organic menu items if customers don't care or can't afford the price difference. Luckily for Kevin, the people dining in Vero's nicest restaurants are willing and able to pay for local and organic. Because their customers will pay more, the chefs will, too. Kevin's spring mix is hours out of the field, while Sysco's or Cheney Brothers' is almost a week out, a fact he doesn't hesitate to use when justifying the price. And of all people, chefs realize that fresh is better. "They love our stuff, and they pay us top dollar for it," Kevin says. "We don't gouge them but we go with what the market will bear, what they will bear. A lot of times the chefs have set the prices themselves and I'm like, 'Okay, that's fine with me!'" These chefs aren't afraid to be demanding in return, however. At one restaurant, a chef specified that the lettuce should be no more than five inches long because the salad bowls are five inches across. Kevin thought his employees were cutting the greens close enough—but the chef, annoyed, pulled out a ruler one day and the lettuce was eight inches long. His salad preppers had to trim it, a waste of valuable time. Kevin encourages chefs to be honest so he can improve. He would rather be humbled by a chef who gets the ruler out than be known around town as the guy who can't deliver five-inch lettuce.

One of these chefs is Executive Chef John Farnsworth, founder of the massive John's Island Club. He's the one who criticizes Kevin most soundly—and Kevin says he can't thank him enough for it. Kevin says John has "schooled" him about what kind of products work in fine dining; he's the chef who convinced Kevin to start growing microgreens, for example. Even though John manages five kitchens and a staff of 127, he personally checks Kevin's produce boxes and catches any and all substitutions, and demands an explanation. Once Kevin substituted organic tomatoes from another local farm because he thought his own had too many blemishes. John ordered Kevin to get his rear end (another word was used in the actual moment) back in the field and bring the Osceola Organic Farm tomatoes instead. I'll cut off the spots, John told Kevin. I'd rather have your tomatoes because they taste fresher, and the customers will be able to tell the difference.

A big part of why Osceola Organic Farm has survived and thrived, then, is that it services a niche market like the one in Vero Beach. Identifying and taking advantage of niche markets, says the Bureau of Labor Statistics (BLS), is a creative way for farmers to survive in a world in which agricultural employment is expected to decline by 19 percent by 2022. As the BLS notes, "An increasing number of small-scale farmers have developed successful market niches that involve personalized, direct contact with their customers. Many are finding opportunities in horticulture and organic food production, which are among the fastest growing segments of agriculture. Others use farmers' markets that cater directly to urban and suburban consumers, allowing the farmers to capture a greater share of consumers' food dollars."[1]

Kevin is doing all of this, and it's working. "We sell out every time," he says about the farmers' market. "We are the only certified organic producer in our county. The market will be dead and we'll have a line." The model works—for now. When and if the area becomes saturated with small organic farmers, they'll have to diversify to compete. All of them can't focus on lettuce like Kevin does. That would be falling victim to the fallacy of composition. For farmers raising commodity crops, the same idea applies: we can't all grow the same crop organically, or prices will fall. Regenerative farmers should diversify, with some growing wheat, others barley, others millet, others oats, and so on. Same with livestock. We need not only regenerative beef, but also poultry, pork, buffalo, lamb, and so forth. All of one's eggs, in other words, should not be placed in a single basket.

There is one unsettling thing about the niche market Kevin has created: it revolves around the upper class. "I'm a farmer, but my job is to service the rich," he says. "That's what I'm doing, whether it's at the farmers' market or delivering to these places. I am a farmer, but it's servicing the rich."

Lots of businesses target the wealthy. But there's a moral difference between food and, say, a Porsche. No offense to Porsche, but no person needs one of their cars to survive. It's becoming more evident, however, that people do need wholesome, regeneratively grown food to live healthy lives. In terms of price, organic and local foods are expensive for most households, and access to those foods is also easier in wealthy urban areas than in poor urban or rural areas. The higher cost and limited availability has led some critics to accuse the organic food movement of being elitist and classist.

I understand that criticism—certainly poor families are at a disadvantage because they can't exercise choice as easily as well-off families, plus they may lack the education that study after study shows is linked to the decision to eat organic. As a 2009 report from the USDA Economic Research Service confirms, "Consumers of all ages, races, and ethnic groups who have higher levels of education are more likely to buy organic products than less-educated consumers."[2] The role of income in that decision, though, is less clear:

> For studies that include income as an explanatory variable, the findings are contradictory. Smaller, higher income households are the

> most likely purchasers of organic produce (Govindasamy and Italia, 1990) and organic apples (Loureiro et al., 2001). One study found that income is unrelated to a household's likelihood of buying organic food (Durham, 2007). A different study found that higher income households are more likely to buy organic vegetables, but once the decision to buy organic has been made, they devote a smaller share of their vegetable expenditures toward organic vegetables (Dettmann and Dimitri, 2010). And yet another study found that income is negatively associated with being an occasional consumer of organic products and has no impact on whether an individual is a frequent consumer of organic products (Zepeda and Li, 2007).[3]

The ERS report finds similar contradictions in the data surrounding the presence of children in the household and notes with a certain level of frustration that "there are no definitive answers about how many consumers buy organic food, how much organic food the typical consumer of organic products purchases, or the demographic profile of the 'typical' consumer of organic products."[4] It seems that higher income does not always correlate with buying organic, but more education usually does.

I understand what it's like to be uneducated about the benefits of organic and too financially strapped to buy it anyway. Fortunately my husband and I can eat organically now, but that was not the case in college or into our twenties. Even if we'd had the money, though, we probably wouldn't have spent it on organic or sustainably grown food because we didn't know we should until I started writing this book. Our childhood communities rejected that kind of food, seeing it as a threat to the livelihoods of local farmers. We didn't learn about its benefits in school or from friends and family. While that's no excuse for our ignorance, I understand how it's possible to know little or nothing about regenerative agriculture (or agriculture in general) and organic food, even in this connected world. Our parents have improved their diets in recent years, but cheap, processed foods were the norm when we were kids. Money was tight, awareness was low, and availability was limited, even for conventional produce. Aside from potatoes and sweet corn, the vegetables we did eat were usually frozen or canned. A testament to this: I bought some fresh portabellas on my way home from the airport a few years ago, purchased from the metropolis of Bismarck, North Dakota, and my mother said with lifted eyebrows, "Are those safe to eat?" She had never encountered a mushroom outside of a glass jar. My husband and I were luckier than many kids, though—we always had enough to eat. When families struggle financially, they tend to grab the cheapest food on the shelf, which is usually highly processed and conventional.

It's easy to criticize the cost of organic food, but that criticism won't hold forever. Organic and regenerative food is not inherently expensive. As Mark Smallwood, executive director of the Rodale Institute, has argued, there aren't enough organic farms right now, which partly explains the price difference between organic and conventional food. "It's because demand is higher than supply," he says. "It's simple economics."[5] Once the organic supply goes up, the price will come down. If we remove the government subsidies that push down the price of conventional food, organic will be even more competitive. And the price gap is already closing rapidly. Organic produce is usually more expensive than conventional, but that's not always the case with processed and packaged foods, which make up most of the American diet. A homemade lunch with organic bread, hummus, cheddar cheese, apples, carrots, and raisins costs less than a conventional Kraft Lunchable, which usually contains crackers, Oscar Meyer ham, American cheese, applesauce, a cookie or candy, and a juice box. Organic yogurt from Butterworks Farm, made with milk from grass-fed cows and sweetened with maple syrup, costs less than Yoplait's Go-Gurt in a tube. On a price-per-ounce basis, packaged conventional food is often more expensive than its organic equivalent that might require a bit more peeling and slicing. Sometimes there's no difference in packaging and the organic option is still cheaper. Conventional Kashi cereal costs more than Nature's Path organic cereal. Uncle Matt's organic orange juice costs less than conventional Odwalla orange juice, a brand owned by Coca-Cola. Upscale conventional pasta sauces cost more than the organic Whole Foods brand sauce.[6]

These comparisons aside, organic food will probably never be quite as cheap, in the limited price-tag sense of the word, as conventional food because conventional food is artificially cheap. As a nation, we've become accustomed to paying relatively little for this food. In 1950 Americans spent 20 percent of their income on food. Today we spend less than 10 percent on food, but that number doesn't take environmental, social, and healthcare costs into consideration.[7] In truth, this industrial food is extremely expensive. If we want to save ourselves from climate change and sickness, then we will have to pay a little more for regenerative food up front, knowing that we are saving soil and water behind the scenes. Though organic agriculture requires fewer inputs, it often requires more labor and management—and people, not machines, must do this labor and be paid for it.[8] And why shouldn't they? Is it so wrong for a farmer and his or her employees, the people who feed us, to make a decent living? We pay professional football coaches, hedge fund managers, and cell phone developers exorbitant amounts of money to do work that isn't as intimate or necessary for human survival as the production of food. If farmers or ranchers can produce healthy, nutritious food, then why not reward them for doing so and thereby incentivize them to continue, instead of punishing them for not cutting corners to make it ever cheaper? And how can farmers and ranchers be expected to feed us if they can barely afford to feed their own families?

Even if society eventually accepts that food should carry a slightly higher price tag, we still need to ensure that middle- and lower-class families can purchase it. The Vero Beach Oceanside Farmers' Market, which started more than five years ago, is one way Kevin reaches middle- and lower-class customers. His organic produce is cheaper or the same price as similar organic produce found in grocery stores. "That's how we get to the regular consumer who doesn't go to these restaurants or these places that we sell to," Kevin says. "We can reach them by going there. That's very ideal to me because when we first started out, I thought, 'I'm going to grow good food for local people.' We started out with that in mind, but we ended up growing good food for the restaurant instead of the people. Now we have the farmers' market. I see rich people there, I see poor people there. We've got a good mix. Even though it's close to the beach, I'd have to say at least half the people come over from the mainland to shop there because that's where the local farms sell and produce companies offer a really good deal there and stuff like that. That's how we reach the whole pie of different types of people."

But not every farmers' market is successful for Kevin. Four weeks in a row he tried a farmers' market in the neighboring city of Fort Pierce and found that customers would not buy his product. The people, he tells me, were not aware of or did not believe in the benefits of organic, so they were unwilling to pay for it. It should be noted that Fort Pierce is an economic and social world away from Vero Beach's barrier island—in the Fort Pierce zip code where the market is held, the median income is $24,071, with just 10 percent of residents holding college degrees.[9] By comparison, the median household income is $102,500 for Vero's 32963, the barrier island zip code where that market is held, and 61 percent of residents there hold college degrees. In both categories, the Vero zip code roughly doubles the nationwide average. Kevin returned from Fort Pierce with coolers of unsold vegetables each week, and finally he gave up on the venture. "If I had landed in Fort Pierce, this couldn't have happened," he says, gesturing across the field. If life hadn't led him to Vero or somewhere just as wealthy, he wouldn't be the type of farmer he is.

Which is sad, because it seems to show that organic farmers and ranchers who don't happen to live near a wealthy urban area will struggle to make it. Or is that just a problem we're facing now, while the agricultural system is still devoted to conventional producers? Perhaps we should look at Fred Kirschenmann's argument for an agriculture of the middle.[10] By this he means an agricultural model that isn't top-heavy, like it is now, but middle-heavy, made up mostly of midsize farms. Farms of this size, he writes, are best equipped to lead the way in replacing conventional agriculture because they are large enough to produce food in real quantities to satisfy food processors and distributors, but small enough to respond to local environments and consumers. They're also small enough to produce unique, highly differentiated products instead of the standardized products we find in grocery stores. Midsize farms also fuel economic activity in rural communities and support families, not corporations.

There's just one problem right now: midsize farms are too big to retail products at most farmers' markets, but too small to compete in the highly consolidated commodity markets. They don't produce enough grain, for example, to sell directly to a mill, but they produce too much to grind and sell at the farmers' market. Kevin's lettuce venture works in the direct market because it is small, but a midsize operation likely wouldn't. That's because food processors and distributors buy from big farms, not midsize farms, for the same reason chefs buy from Sysco and not individual farmers—it's too much of a hassle. We need to reorient our food system so that regenerative midsize farmers can market their products locally and regionally, rather than participating in national or world commodity markets. If we do this, then responsible farmers, no matter where they are, will be able to enjoy the economic benefits of regenerative production.

But we don't need a complete transition to an agriculture of the middle. Our ideal should be diversity, just as nature's ideal is diversity. The goal isn't to create a nation filled with one size or type of farmer, but a nation of diverse farms that includes many more midsize farms than we have now, as well as large operations like Phil's and small ones like Kevin's when the environment and market conditions dictate. Phil's big ranch works because it fits with his goal of restoring as much native prairie as possible. Likewise, Kevin's farm serves a small, but not unimportant, market, which is his community. While it's nice to market directly to consumers, that's not possible—and not desirable—for every type of operation. There is no one-size-fits-all model for marketing, just like there is no one-size-fits-all model for what crops and livestock should be grown where or who should do the farming.

Our next stop is Osceola Bistro. Again, we enter via the back door, right into the kitchen. Employees in white uniforms season bright red racks of ribs in stainless-steel roasting pans. I've never smelled a room so delicious, though I don't know what exactly I'm smelling. Kevin knows everyone—he gives high-fives, shakes hands, folds people into hugs, chats with them. He's like a celebrity. We take a few moments to chat with chef-owner Christopher Bireley, who praises the freshness and taste of Kevin's produce as well as the personal service. After we leave, Kevin tells me that he refuses to sacrifice that kind of personal service for the sake of opening more restaurant accounts. "I turn down so many people every year because we can only do so much with ten acres and the time we have," he says.

This is curious to me—Kevin knows there are customers waiting for his product, but he has not expanded. Capitalist thinking suggests that more acreage would allow him to capture more profit. Why haven't you gone down that road? I ask. Part of the reason is that expansion would mean handing more responsibility to others, and he hasn't found the right person for that. Expansion also does not guarantee more money or market share, and he's skeptical of partnerships and debt. "We've done everything out of pocket," he says. "I don't go to a bank every year and borrow money to get my farm going and stuff like that. I guess it's sort of old-fashioned, not really wanting to go borrow money to get bigger. Expanding is bigger, but that doesn't mean better."

But the major reason Kevin hasn't expanded is that money is not what drives him to farm. "I feel like I work enough, and I always want to have time for my family and to go surfing and fishing a little bit. That's very, very important to me," he says. "I don't know if making more money would change my life. I feel like I make enough money that we're happy. That's the bottom line, to be honest with you. You see people that got a lot of money and all they do is worry about their money. I worry about if I have time to go surfing or be on the beach for a little while this week or something like that. There's a balance. We all have to have money and be fiscally responsible and all that stuff. But being rich isn't a goal of mine. I've never had that ego to push me to get bigger to make more money. I've just never been that way."

Agribusiness supporters would likely scoff at Kevin's admission. They would say that one farmer loading a car full of produce a few times a week is neither success nor progress. What Kevin does, they might argue, represents a step backward, a return to the "old days." He hasn't specialized, mechanized, or corporatized. He uses little technology and oil. He doesn't participate in the farm program. He owns one small tractor and ten measly acres. He employs local workers, like retired nurses who live up the road, and pays them a living wage instead of hiring cheap migrant labor to reduce his cost of production. Under the "get big or get out" mindset, Kevin's farm is too small and too unproductive. There's no place for him in modern agriculture except as an outlier.

Yet Osceola Organic Farm's very existence proves that thinking wrong. Kevin doesn't need to be big to be profitable, produce quality food, and enjoy his work. He's making more money with less labor and debt-induced stress than a lot of industrial farmers are with their one-thousand-acre-plus farms and fancy machinery. He is living proof that human beings don't have to degrade the land to eat. Between him and the average conventional farmer, I can see who is better off in the long run.

This is what's coolest to me, though: Kevin isn't doing something that the average American couldn't also do with a little start-up capital, research, perseverance, and a real desire to produce healthy food and preserve the soil. He doesn't have generations of farmers in his family tree, nor did he start farming as a young man. He had never grown food for anyone but his family before buying those ten acres. He switched careers in the middle of his life. Kevin's story shows that almost anyone can be a farmer if he or she truly wants to. Given his example, maybe we should offer farming as a respectable, desirable career option for young people, just like we offer nursing, banking, or engineering. What would happen if college or high school graduates could enter educational programs to become regenerative farmers or ranchers? What if young people saw farming as socially acceptable, as acceptable as it is to enroll in a university? Right now, we have a few farmer-training programs scattered across the country, sponsored by small local organizations and state governments. What if these programs were more common and served more people? What if they drew young people from crowded urban areas to rural places to revitalize those communities?

What if these programs welcomed existing farmers, too? We're almost one hundred years out from the days when regenerative farming lessons were taught on the farm. Only conventional farming is passed down now. Even in our internet age, learning how to grow crops and livestock organically and regeneratively in specific environments isn't easy—and as any good farmer knows, *every* environment is specific and challenging, as nature intended it to be. Replacing generations of industrial lessons will take training, in part through hands-on learning programs, classroom courses, and mentorships. This training can't be based on one-size-fits-all thinking—we've seen the danger of that mentality—but should be designed for specific environments, markets, and social conditions. We have a decent start already, with more universities offering classes in organic agriculture, more organizations popping up that are dedicated to assisting organic farmers, and more information on sustainable agriculture appearing on bookstore shelves. But much more can be done.

Some might argue that it's not fair to ask farmers and ranchers to trust off-farm knowledge. That's one reason conventional farming took root: farmers let agribusiness "experts" dictate their practices. I agree that we shouldn't ask farmers to forsake their autonomy in the transition from conventional to regenerative. But who says this knowledge has to come strictly from off-farm sources? We could empower farmers like Phil and Kevin to educate their communities. A neighbor-to-neighbor approach seems better than a top-down, prescriptive approach, anyway. Some farmers and ranchers will find structured training helpful, but others will want more independence. And not all farmers are going to want to make the transition—and they shouldn't be forced to. But they could be encouraged in the same way they were encouraged to grow more corn, adopt industrial practices, and "get big or get out." If society is going to ask for a transition, then society must also provide resources to help make the transition happen.

We drive by Kevin's house on the island. Two blocks from his front door, the Atlantic Ocean that he loves so much laps at the sand. "I feel really blessed," he says, looking out the window. "I love what I'm doing. There are some days that are tougher than others, some days that don't work like they should, some days go perfect. We have a lot more good days than bad days."

"I'm very lucky to be where I am," he tells me later. "I'm in a unique position. I grow food that keeps me healthy, keeps me young. I'm doing that work, and it's good exercise in fresh air. I couldn't be more blessed."

17
The Urban Farmer

I am a country girl at heart. Living in the city can be fun, especially in sunny South Florida, but most of the time I feel boxed in, like I can't quite breathe. I miss my parents' ranch: thousand-acre prairie pastures with nothing but cows, antelope, deer, grass, sky, and the occasional barbed-wire fence. No people, except me riding my paint mare. No sounds, except the wind in my ears, the chirping of meadowlarks, the rustle of grass. On weekends my husband and I flee in our Jeep with a tent and a cooler of beer for the solitude of places like the Everglades, Ocala National Forest, or Florida's wildlife management areas. But our jobs require a city, so for now we stay.

Seeing Kevin's farm, though, cracks open a door of hope. He's proof that perhaps I, too, could go back to the land, even if it's only ten acres on the outskirts of town. Surely other people feel unfulfilled in the city and want to get their hands in the dirt. Maybe like Kevin, we've only grown organically in our backyard garden, maybe we didn't grow up on a farm, maybe we'd be taking a financial risk, maybe we can't stray too far from the city. But maybe we feel it's time to return to the land in some way. How, then, do people like us get started?

I'm sitting on a plastic lawn chair in front of a small, red-brick house. A yellow lab puppy rubs his face into my pant leg and I instinctively scratch his back, not realizing I'm doing it until he wanders off and my hand is left dangling. Two grown labs, one chocolate brown and one black, doze in a patch of sun. The yard in which my chair is located is mostly hard-packed dirt—this is the desert, after all—yet the gnarled cottonwood tree nearby seems unconcerned. It stands like an ancient statue with its leafy arms spread over the house and the street beyond. In fact, yards all over this neighborhood boast photo-worthy mature trees like this one, their roots sucking water from the Rio Grande riverbed below.

The Rio Grande—the most romantic-sounding river name—is just across the street, hidden by more trees and the built-up bank. I hear geese chatting about who-knows-what. When I crossed the river in a taxi earlier, the water looked like melted chocolate ice cream and was moving just as slowly, muddy and sluggish. It is a February afternoon and the temperature is in the fifties. No clouds, just stark sunshine glaring off the brown yard. In the sun it is warm—I get a light sunburn on my chest, just below my neck—but when a shadow creeps over my chair, I reach for my coat. This is my first time in the American Southwest. Officially, I'm presenting an academic paper at the Southwest Popular Culture/American Culture Conference in Albuquerque's sterile, quiet downtown. Unofficially, I'm doing research for this book: interviewing Fidel Gonzalez, an organic farmer making his living just outside the city limits. I'm here to tour his backyard greenhouses and learn what it's like to be an urban farmer.

I found Fidel by contacting the American Friends Service Committee (AFSC) office in New Mexico.[1] AFSC operates a farmer-to-farmer training program under the leadership of codirector Don Bustos.[2] Bustos and others provide personal, hands-on training to beginning farmers for one year. The participants learn about crop selection for desert environments, soil health, irrigation, harvesting, food-handling techniques, business planning, and marketing. They are students of the earth for 365 days and are then released into the wild that is running one's own organic farm. Thankfully they can call their mentors anytime. The program also provides economic assistance in the form of tools, seeds, equipment, and other as-needed items. No agricultural experience is necessary, only the will to succeed.

Fidel, forty-seven, graduated from the program and has farmed organically for the last five years, a job that literally and financially feeds him, his wife, and young son. Like Kevin, Fidel looks nothing like a typical farmer. No John Deere green in sight. He wears a straw hat with a black hatband, a black-and-white plaid jacket, and a yellow T-shirt underneath with a funky Aztec-inspired sun-and-moon graphic. Beaded necklaces lie against the collar, and I spot a flower tattooed on his hand. His black hair is clavicle-length and loose. He wears a thin, close-fitting silver nose ring, two silver earrings, and round, gold-framed glasses à la Elton John. He looks . . . ruggedly hip.

Fidel was born and raised in Mexico City and moved to the U.S. at twenty-two years old. Before going into agriculture, he was a professional musician. When he moved north of the border, he jammed with South American musicians in New York City and they formed a band. They played at the Gathering of Nations Powwow, the country's largest Native American cultural festival, at the Atlanta Olympic Games, and all over the country, more than a decade of windshield time. Fidel lived on the road, always performing, always moving. On tour once, the band stopped in New Mexico. "I fell in love with New Mexico," Fidel says. "I was trying to come back every year to play."

In 2000 an exhausted Fidel realized that he did not want to be on the road so much, that he wanted to stay in touch with music but needed to put down roots. He hadn't married because he couldn't stay in one place long enough to date. Later, he moved to New Mexico, where he set up a home recording studio and started producing other people's music. But he couldn't generate enough income. He took odd jobs. His bank accounts dried up. In need of cash, he helped make a documentary with a friend, Pablo Lopez, who was also working with Agri-Cultura Network, a farmer-owned produce brokerage that markets certified organic, local produce from Albuquerque to restaurants, public institutions, and farmers' markets. Fidel asked Pablo if he knew about any open jobs, anywhere, doing anything. Pablo suggested going into agriculture through the AFSC farmer-to-farmer training program. Maybe Fidel could join Agri-Cultura Network someday and market his produce through the organization, Pablo added.

"I had never done anything in my life with agriculture," Fidel says. "When Pablo talked to me about greenhouses, drip systems, drip irrigation, cover-cropping, he was talking to me in a different language. Words I had never heard in my life." Like most Americans, Fidel was a city dweller, a man of asphalt and concrete, not soil. He'd never considered farming as a career. Financial desperation fueled his decision more than a commitment to organic and sustainable living and eating, a commitment he has now but didn't then. "At that time, it was more about saving my skin from being kicked out of the house where I lived," he says. "So I told him yes. Five years later, I am here running my own business." He smiles and gestures toward the greenhouse behind the brick house. The sun glints off his nose ring. A rocker turned farmer.

Fidel's farm is USDA certified organic, with nothing that will hurt the earth, the crops, or humans, he explains. He grows lettuce, arugula, carrots, tomatoes, radishes, chard, kale, and collard greens, with plans for more varieties of herbs and vegetables. "Lettuce is all over," he says proudly. Blood meal and bone meal, both approved biological additives, provide most of the soil's supplemental nutrients. "Blood meal and bone meal are welcome in the organic world, unless you want to be totally vegan," Fidel says. The greenhouses, not pesticides, control insects. Instead of herbicides, crop rotations effectively keep most plant diseases and weeds in check, with hand-weeding as the final defense.

Fidel irrigates the crops using a system of drip lines, much like Kevin does. "Traditionally, here in New Mexico, one of the techniques that they used to use for watering was the ditches that overflowed on the land," he says. Indigenous people living along the Rio Grande dug irrigation ditches that date back to the 1500s.[3] These ditches carried water to corn and vegetable fields, allowing human life to flourish in the desert. "The drip lines are a small version of that," Fidel says. "The idea is to preserve energy in general, and water is one energy source. The drip system really helps us because we save lots of water. We just water the land that we want to get water, and you don't waste energy cleaning the piece of land that you didn't plant and all you have are weeds."

Water stewardship is more important than ever. Drought has plagued New Mexico over the last three years, Fidel tells me. The previous two years were the driest ever recorded, and the region in general is warmer than it was fifty years ago. Groundwater supplies have dwindled, and some farmers along the Rio Grande don't have enough water to irrigate. Dust storms sweep through, eerily reminiscent of the Dust Bowl.[4] The entire Southwest, in fact, has experienced drought, which is only predicted to worsen: a 2014 study indicates an 80 percent chance (90 percent in some areas) of a decade-long drought in the American Southwest and, depending on how much climate change worsens, a 20 percent to 50 percent chance of a megadrought, or one that lasts thirty-five years or more.[5] Even a decade-long drought, short compared to thirty-five years, hasn't hit the U.S. since the Dust Bowl—and with the return of farming practices from that era, such as continuous cropping and massive fields with few fencerows and shelterbelts, we can expect such a drought to have a devastating effect on agriculture. Meanwhile, sprinklers keep golf courses and front lawns green and plush across the Southwest.

The problems New Mexico faces are actually nationwide problems, just amplified. In every state, industrial agricultural exacerbates water problems, drought or no drought. The environmental effects of agricultural runoff—fish and amphibian deaths, algae blooms, plant die-offs—only worsen in sources that contain fewer gallons of water and are therefore more concentrated with agrochemicals. Yet the Southwest is at the epicenter of what is already a water crisis, which could by default become an agricultural crisis in which the region produces far less food. In a pre-climate-change world, an extended drought in a single region and the collapse of its agricultural sector would be relatively manageable in terms of its effect on the national food market. Other states could compensate for reduced agricultural production in the Southwest. But we no longer hold such an insurance policy. The future of agriculture is shaky all over the United States as the effects of climate change reveal themselves: rising temperatures that cause crop failures, water shortages in some areas and severe floods in others, increased carbon dioxide levels that reduce the nutritional quality of pastures, and desertification. Very few farms now produce the majority of the nation's food supply—this lack of diversity means that if anything happens to those big farms because of climate change, we are in trouble.

The question of who gets water is at the center of the climate-change discussion. Setting water aside for agricultural use is a good and necessary thing, as long as the users operate sustainably. As a society, we shouldn't allow the careless use of water by giant, environmentally disastrous CAFOs, but we should encourage the reasonable use of water by carefully managed, grass-centered ranches. We shouldn't stand by as industrial farmers fill eight-hundred-gallon tanks with water and pesticides, but we should applaud as organic farmers increase the carbon levels of their soils so that they absorb and hold water better than conventionally managed soil. For now, though, all farmers fight for a share of the water pie, and it's a battle that grows fiercer with each year of drought.

At least water goes to good use in Fidel's greenhouses: it helps feed people, mostly schoolchildren, nutritious and chemical-free produce. Still, he's careful to conserve. "We have to be extra careful with water," he says. "The idea of farming using the drip system is precisely about how to take care of the water we have as a community. We are in the middle of the desert." The dry air chafes the inside of my nostrils with every breath as we sit and talk. I remember the airplane's descent into Albuquerque: the brown that extended as far as I could see, crisscrossed by slightly lighter brown roads. The listless Rio Grande is across the street. We are in the desert, that's for sure. One growing hotter and drier by the year.

Living where one works is the blessing and curse of being a farmer and rancher. There's the physical beauty just outside the farmer's doorstep: the hayfield shimmering with dew, the yeasty smell of grain inside a bin, a pasture dotted with cattle. By inhabiting the land, the farmer experiences a oneness with it, a sense of being pleasantly caught in its rhythms. These emotions can lead to feelings of completeness, of deep fulfillment, of having been called to this particular occupation.

But there is also the unending work visible through the kitchen window and audible in the bawl of a calf in the corral. When things are stressful—if there's a drought and everything is dying, or if a blizzard has dumped so much snow that just getting from the house to the barn is a challenge—then there is no escaping the stress. On conventional farms, living where one works can even be a health hazard. Agrochemicals drift from the field to the farmhouse and seep into the groundwater that feeds the farmer's private well. The stench from hog or chicken confinements or cattle feedlots ruins air quality and pollutes water sources, and makes children struggle to breathe and suffer skin rashes.

Life on my family's conventional ranch is just such a mixture. The ranch's towering flat-top hills, lazy creeks, rolling grasslands, and tree-lined gulches provide picturesque scenery. The prairie is all solitude and stillness, grand vistas and open skies. Some people feel exposed on the prairie, unable to hide; not everyone feels comfortable there, and I like that I do. It's a place whose beauty never fails to inspire in me a sense of being very close to God. But all this falls away when I smell Roundup on the evening breeze as I sit on a lone hay bale watching the sun go down, or when I see the arrow-straight rows of GM corn, their brown tassels sharp against the blue August sky. Now I fear the water running from the kitchen tap, the beef on the table. One *is* exposed here—there's nowhere to hide from the toxic effects of conventional agriculture.

To say Fidel lives where he works is an understatement. Half of his farm is literally in his backyard, closer to the house than any farm I've seen. He spends his days in two thirty-by-ninety-six-foot greenhouses, which together comprise the total growing space on his urban farm (he has plans to build two more on rented land roughly ten minutes away). One greenhouse stands behind his house and the other fills a neighbor's backyard a couple of houses down. The farm also includes a brick side building for storage and starting seedlings, a walk-in cooler, and a shed. A waist-high iron fence separates the greenhouse from the rest of the yard so that the dogs cannot dig up or contaminate the produce. The greenhouse rises like a cathedral, and the sunshine makes its white plastic appear to glow from the inside out. A wood-and-metal frame holds the rounded structure together. Seven long beds stretch from end to end, filled with lettuce, arugula, and radishes. One bed is heated using solar panels installed by a local university, perfect for cold desert nights that might harm the plants. Squatting down, he shows me some lettuce mix that will be perfect for harvest on Monday.

Fidel markets his produce in four primary ways: through the city's public schools, through restaurants, at farmers' markets, and via a community-supported agriculture program that provides boxes of produce to member families.[6] Agri-Cultura Network is the go-between for all transactions except for what's sold at the farmers' markets, which Fidel handles himself. He works two markets a week, one in downtown Albuquerque and another about ten minutes from his house. About 50 percent of what he grows ends up in restaurants, CSA boxes, and the farmers' market stand. The rest lands on the lunch trays of school children (the public schools purchase only salad right now). "They are supporting the local agriculture and feeding the kids healthy produce," Fidel says. "They haven't changed completely—they still give them hamburgers and stuff like that—but this is a step ahead. It's a really big step to feed the kids healthy vegetables and salad greens with no chemicals and stuff like that, and to support the agricultural movement right here in New Mexico. That will keep the money here with us and make the economy stronger." It also means that Fidel's farm doesn't primarily serve the rich, but the entire Albuquerque community.

Fidel's days now revolve around the growing cycles of arugula and carrots instead of the cycle of summer music festivals. His fingers pluck weeds out of the lettuce more often than they pluck guitar strings. He sells produce at farmers' markets instead of performing at them. He recalls how he used to strum away next to the produce stands and see that people enjoyed the music—but they spent money on food, not CDs. Food is the first human need, he realized. "Music is great! But with music I cannot take your hunger away," he says. And this is a hungry community. According to the U.S. Census Bureau, almost 22 percent of New Mexico residents live in poverty.[7] Only Mississippi's poverty rate is higher. Fidel feels good about the fact that public school kids, no matter their socioeconomic status, eat his organic lettuce for lunch. But not everyone can afford what he sells at the farmers' market. He sees that when families make food decisions, money often drives the choices. "It's hard when there are families with five, six, or seven kids that they have to feed and they are thinking, 'One pound of lettuce at the farmers' market is eight dollars organic, versus two-something at Wal-Mart.' In a moment like that, I do not judge anyone because I am not in their shoes," he says.

Fidel understands the fraught connection between food and poverty. He says the best thing he can do as a farmer is educate people about the benefits of organic and regenerative food and the artificial cheapness of conventional food. He tries to "lead others to change their consciousness" and to "wake people up rather than ignore the problems" that industrial food creates. The cliché that change begins with one person suddenly takes on significance as Fidel describes his vision for creating a regenerative agricultural future. He leans forward in his chair, elbows on his knees and palms pressed together as if he's praying. "We have to have a sustainable community," he says. "I believe that in order to have a sustainable community, you have to begin with you." He points a finger at his own chest. "You have to be a sustainable person yourself. Then you will be able to have a sustainable family. Then you will be able to have a sustainable community. Then you will be able to have a sustainable society.

Then we will be able to have a sustainable country. But the whole entire thing begins one person at a time, us. It's a consciousness movement."

Changing people's consciousness about food and agriculture, he says, will also help expose the heavy influence of conventional marketing. "Before the industrial movement, before being called organic or pesticide-free or chemical-free, that's the way everything used to be before. Anyone could have it at their tables. It was so easy before industrialization," Fidel says. "Everyone could eat just the way we try to eat right now, healthy. Then they came with all the changes, all the propaganda of, 'Eat this and you are going to look better, smoke that and you are going to have class, drive this kind of car and you are going to feel like a king.' All that materialism, we can't deny it is here. Now the problem is, how does that affect us to such a point that people start to feel comfortable just with fast food? So it's not all about classism, it's about how people want to live their lives. Like I said, I understand if there is a family with more than five kids they have to support. But at the same time the leadership of that family, the father and mother, can understand that instead of spending money on something they do not need, they can feed the children properly so they don't have to worry about diseases later. We know it's expensive at the time, but when you see over time the benefits of it, in reality it's not expensive at all."

Given his Mexican heritage, Fidel feels a special responsibility to reach out to Albuquerque's Latinx community that makes up 48 percent of the city's population.[8] Though they're half the city, Latinx families are overwhelmingly the minority when it comes to farmers' market attendance, Fidel says. He wants to change that. "Of 100 percent, about 75 percent are women. But you see all kinds of people. Unfortunately you don't see too many who speak Spanish, Latinos, buying organic food. It's something we need to work more with, education, to educate our people," he says.

The conversation takes a philosophical turn (I'm noticing a trend with my interviews for this book: a return of the farmer-philosopher of old, the yeoman). Fidel sees the lack of education about food—and the lack of wholesome, nutritious food available to all regardless of socioeconomic status—as an injustice. But injustice is everywhere, he points out, so how do we decide which of the many forms of injustice to fight? Fidel mentions the wars of the world and the drug-fueled chaos in Mexico City, his hometown. He muses that he could have been a soldier and fought injustice that way. But he doesn't think violence is the best use of our energy. "How am I going to be more beneficial for the human race, with a rifle in my hands or with seeds in my hands? I respect what people are doing in places because they have to. There are no choices sometimes. It's what you have," he says. "But thankfully right here with us we have the opportunity to choose what we can do to help. And I don't see anything else better than agriculture."

18

The Agriculturalized City

Fidel, his wife, Raquel, and their young son, Tenoch, returned three weeks ago from a visit to Mexico, where they stayed with Fidel's family and toured the ancient pyramids at Teotihuacán. Fidel tries to visit Mexico at least once a year. Raquel is originally from Puerto Rico, so the family travels there as well. Fidel's dream is to take his family to India to see the Himalayas. I'm not surprised by this dream; Fidel is a worldly man, a person who understands the importance of appreciating and learning from other cultures. He is an artist at heart, and the beauty of India's landscape, food, and music appeal to him much like the vibrant culture of New Mexico did years ago. Traveling is most difficult for him during the peak growing season, which is during the school year. The greenhouses allow him to grow year-round, though, unlike most New Mexican farmers. The summer heat can be brutal, but he says proper attention and preventative steps like growing drought-resistant crops and providing extra water or shade can mitigate those problems. If he stops growing, it's only because he wants to take a break during the slower months.

Traveling is something Fidel can better afford now that his business is generating profit. He survived the three-year waiting period that comes with organic certification, and every year his crops are more productive as his knowledge grows. "The first three years, it was a challenge for me," he says. "As far as I understood when they were training me, they told me right away, 'If you put good energy into what you are doing, you are going to see results in the next four or five years. It's going to take you awhile. You are not going to see money right away. You're going to make money enough to pay your bills, but a good profit is going to take four or five years.' This is going to be my fifth. So by now I can tell you that last year I had really good production and I expect this year is going to be more than that. I can just tell you by my experience, even if it's a short experience, it's a beautiful way of living economically speaking if you pay attention to what you have to do. Agriculture in general is a good way of life."

The farmer-to-farmer training program sponsored by the American Friends Service Committee purchased most of what Fidel grew that first year so that he had a consistent market. The program also helped pay for the greenhouses, a rototiller, and a raking machine, and mentors like Bustos provide continuing support. Practical farming knowledge, Fidel says, was the most valuable thing he gained from the program and his mentors. They didn't just swoop in and set him up with equipment and seeds, then disappear. The program lasted a full year. The training was hands-on and personal. Students and teachers stay in touch, see each other around. The program also created a community of like-minded regenerative farmers, an invaluable resource in a world where agriculture classes and resources overwhelmingly service conventional farmers. I'm reminded of the Chinese proverb, "Give a man a fish and you feed him for a day. Teach a man to fish and you feed him for a lifetime." The program taught Fidel and others how to fish for the long-term.

But the greatest gift of all came later: personal fulfillment. Farming was just a job at first, he says. It's way more than that now. "Now I wake up every day and I feel good going out working in the greenhouse, preparing the seeds. I feel that satisfaction when I see my son eating what I plant," he says. "I remember the first time we gave him chard and collard greens. When I see my family eating what I plant, I feel the satisfaction."

His days are split two ways now: agriculture in the morning, family in the afternoon and evening. "I want to be there with my son and watch him grow," he says. A typical day starts with farm work from about 6:00 a.m. to 1:00 or 2:00 p.m. Fidel takes a break, then plays with his son, practices for upcoming shows, or works on recordings. He performs a music gig once a month or so. He wouldn't trade farming for the musician's life now, he says, even though music is still an important part of his identity. He's found another way of feeling alive. "You need to touch the soil, you need to smell it," he says. "Sometimes when we are harvesting, I just grab the carrots—for some reason I like the mix of the carrots and soil, maybe because of the sweetness of the carrots and how they smell—and you feel like, 'Wow, this is *life* for real, man.' I'm not saying what other people do is not life, it's just that it means more for me what I'm doing now than what I used to do before."

This is what's been lost for many because of industrial farming, what farmers and ranchers desperately need to feel again: a connection with the land. That this is *life* for real, man. Industrial farming demands that farmers set aside feelings and focus on numbers, profit, yield per acre. This approach makes farmers simple producers, not stewards. People become machines: replaceable by other machines, dispensable, their job secure only as long as they are mechanically necessary.[1] They are bodies inside tractors that inject GM seeds into chemically sterilized soil. They are not men and women who feel joy at the smell of carrots and soil mixed together.

I tell Fidel it sounds like he sees agriculture as a spiritual experience. He says yes, that agriculture is part of what it means to be human. "I do see agriculture as a way of life, a way of living," he says. "Agriculture means everything in life. Everything comes from agriculture. We are talking about social, political, artistic, religious, whatever, anything." Having food allows a society to create art, contemplate philosophy, make music, discuss politics, things that aren't directly connected to physical survival. I hear an echo of Phil in Fidel's words. *If we don't do our job right, there isn't room for others.*

Fidel's three labs bark at people who ride by on horseback or stroll past on foot. The neighborhood is silent except for these dogs, and their barks echo down the street. The yellow puppy barks at everything: his shadow, the toys in the yard, me if I stand up too quickly. Downtown Albuquerque is less than ten minutes away by car, but sitting under the ancient cottonwood tree, I would never know that if I hadn't passed through on the ride over. This side of the river, a thirty-nine-square-mile neighborhood called the South Valley, is outside the city limits, but with its streets and houses, it's clearly not the country. Tiny one-hundred-year-old adobe houses still exist here, but so do fancy equestrian parks. Not urban, but not rural. A local real estate website calls it "rurban."[2] The lack of land for sale in the South Valley is an issue for Fidel. "The land is not even our land," he says, waving toward his second greenhouse down the street. "We have friends who allow us to use their land. Because we are limited in that part, we are trying to figure out how to make it productive. Ideally the best thing would be to use it one year and let it rest one year. But that can only happen if you have a good quantity of land."

By resting the land he means planting cover crops to replenish the nutrients used by the cash crops, like Kevin does. With only 5,760 square feet of land, however, Fidel can't afford to devote any to covers, so he needs to expand. The planned addition of two greenhouses is a start, but he'll still lack the security and flexibility of land ownership. Bustos told him that three acres of land, when managed properly, can produce enough food and income to provide for a family. "So we need double that to have the other half in cover crop, so we need six to ten acres," Fidel continues. "That way you will be sure you are taking care of the soil and the earth while you are doing business, too."

"Having land is not about being a more powerful farmer or whatever, it's about creating the community," Fidel continues. "Having our own land, a good piece of land, means producing more jobs, too." Fidel recognizes—and I do, too—that he has the potential to lead people, to be a job creator, to use agriculture to improve people's lives not only in terms of their health, but also their career. Expanding production would allow Fidel to employ locals as Kevin does. "Let's say I could use somebody else's land, another acre. That means I can produce a job for another two or three families," he says. "We produce healthy food, we make the economy stronger in New Mexico, and we produce local jobs." Fidel's "we" means the state's organic and regenerative farmers.

In a desert the most secure place to farm is near a water source. Though the Rio Grande is at record-low levels, it's still one of the most stable water sources in the region, so Fidel is hoping to acquire land near it. He also wants to stay close to his urban customers, keeping the transportation footprint of his produce minimal. "If I can find something like this"—he points at his backyard—"no more than thirty minutes away that would be great," he says. Fidel was quoted $105,000 per acre for a piece of land near the river, not far from his house—an exorbitant cost compared to what other farmers pay. Farmland in Iowa costs, on average, $8,750 an acre, according to the USDA. Farmers in New Jersey pay $13,000, the highest in the nation. The U.S. average is $4,100. Other New Mexico farmers shell out just $1,450 an acre—the second-lowest nationally.[3] But Fidel isn't trying to buy farmland, he's looking at prime riverside property that could be developed into expensive homes. At the quoted price, he would pay more than $1 million for ten acres. He doesn't have that kind of money, and it's unlikely that he ever will. His options are either to finance a large purchase or buy small pieces over time. Even if he abandons the river idea, staying within thirty minutes of his house means he'll still pay Albuquerque prices. This is the challenge urban and rural farmers alike face if they don't inherit farmland—and with the nation's farmland consolidating into fewer hands every year, the challenge only grows more difficult.

Fidel isn't letting land woes get him down, though. Overall, he says, his worries are few. Despite being a city, Albuquerque is not a bad location for growing food: no hailstorms, tornadoes, major floods, or long winters. The economic and social environment is even better. "It is a good thing to be a farmer now in these times in New Mexico," Fidel tells me. "We have had really good reception from the community." Fidel is right: it *is* a good time to be a farmer, particularly in Albuquerque. The city is about to undergo a quiet agricultural revolution. Farmers like Fidel are part of the city's larger goal of reducing its greenhouse gas emissions by 80 percent by 2050. Albuquerque's Climate Action Task Force has proposed an eight-part plan, pending city council approval, to achieve reductions.[4] One part involves local food and agriculture.[5] This piece is further divided into four strategies: (1) increase the amount of food produced inside city limits; (2) support the development of the food shed in New Mexico ("food shed" means the geographic region that produces food for a specific group of people, the goal being to encourage New Mexicans to buy within Albuquerque's food shed, or within a three-hundred-mile radius of the city); (3) incorporate food and agriculture in city planning, landscaping, and design; and (4) engage every city department in promoting local food production and consumption. The proposal is chock-full of concrete ideas: gardening classes for children and seniors, incentives for private and commercial greenhouse farming, land to be set aside for community gardens, edible landscaping, bus service to farmers' markets, and the establishment of a "buy New Mexico" preference for city food purchases. Urban farmers like Fidel represent a new component of Albuquerque that the city wants to nurture in order to achieve an energy savings target of 25 percent from 2000 levels by 2020. In short, the city wants to encourage agriculture and revolutionize the food supply.

I look at Fidel and see the future that Albuquerque's climate task force must have envisioned: a city agriculturalized from the inside out. A diverse collection of farmers and livestock producers working within city limits and beyond to feed people in a way they haven't been for more than a century: locally, regeneratively, and thoughtfully. The city's carbon footprint drastically reduced. People armed starting in childhood with the knowledge to feed themselves, people supplementing their grocery lists with produce from community gardens, edible landscaping, and backyard gardens. It's an inclusive future, one that welcomes young and old, rich and poor, service workers and wealthy business owners. This is what we need nationwide: regenerative agriculture happening in the city as well as the country as part of a broader commitment to responsible living.

Fidel and I climb into his small, battered Ford pickup, similar in size to a Chevy S–10. He's driving me back to my hotel for the night; tomorrow I will present my paper. We cross the wide Rio Grande. I ask how he responds to those who might say his farm is too small, doesn't feed enough people, isn't a real, honest-to-God farm. Making a difference is a matter of perspective, he says. "Some people will say, 'This is nothing compared to what we have.' Some other people will say, 'I wish I could have at least a ten-by-ten planted! How do you put the seeds in? How do you do this?' It's all relative," he says.

He opens a bar of dark chocolate with almonds and Stevia. He offers me some and I snap off a square of the pleasantly bitter, nutty chocolate. Fidel, a trim man, says he is trying not to eat so many sweets (the opposite of my approach to them). He was actually prediabetic at one point, but that went away after he changed his diet to mostly organic food. As we munch the chocolate, Fidel points out that you can never make everyone happy. Half the world thinks the Dalai Lama is a saint and the other half thinks he's a devil, he points out as an example. Same with the pope or the Aztecs of old. There is no way to make everyone agree on what is right and good, he tells me. So how do *you* know what is right and good for you, I ask. Fidel pauses, then says, "Your heart is always going to tell you the truth. When I see my son eating, when I see my wife smiling, when I see my friends happy when I give them some lettuce, in my heart I know I am doing the right thing."

PART FOUR
Diversified Regenerative

19

The Diversified Farm

Home. I go the first week in August, when wild sunflowers bloom in the road ditches and the warm bread smell of wheat fills the air as combines chug across the fields, spewing straw behind them. Golden grain pours into the chalk-gray bins, and my father keeps some in a coffee can in the pickup, partly for taking to the elevator to test the protein and moisture but also for snacking. I ride my paint horse through the pastures and yellow sweet clover brushes my boots in the stirrups. Sometimes I'm too lazy to put the saddle on, and soon my Levis smell like horse sweat and are covered in white and brown hair.

I walk alone into a hayfield one evening with my camera and a beer and climb onto a round hay bale. It takes a running jump and an awkward scramble, and the hay leaves red scratches on my forearms and shins. When I was a kid, the run-jump onto bales was as effortless as stepping over a puddle. The hay is still warm from the afternoon sun. A doe and her fawn leap over a barbed-wire fence and bound into the pasture on the other side. Coyotes howl. I photograph the sunset and drink my beer and cry a little because it's been so long since I've felt the peace, the belonging, the fierce but loving hug of the prairie. I haven't been home in the summertime for two years.

I spend a lot of time with my brother, Joshua, on this trip. We lurch across the pasture in the pickup to check the cows and the water tanks, go 'round and 'round the hayfield in the swather (I drive partly for fun and partly to prove to myself that I remember how), and practice jitterbugging in the living room so he can impress the girls at a Saturday night dance. One morning he makes French toast from bread he baked himself. He wears his hair long for around here—it just brushes the top of his shoulders—and he's one of the kindest, most patient people in the world. We talk about holistic management, organic farming, and the pitfalls of industrial agriculture. Josh is interested. He's always been interested, but now he's serious. He wants to get started running some grass-fed, organic cattle, maybe next summer.

Josh is twenty-one years old. He didn't go to college. Didn't like being in the classroom, didn't see the point when he could learn how to farm and ranch from Dad. The problem is that he's not learning the important stuff by working on our conventional operation, at least to my mind (Josh is beginning to think the same thing). I ask if he knows what it means for plants to fix nitrogen. He doesn't. What about cool- and warm-season grasses—what's the difference? Not sure. I sprinkle other questions in during the week: What nutrients does corn take from the soil as opposed to small grains like wheat or oats? What's the purpose of a cover crop? Why is it important to have carbon and organic matter in the soil? He doesn't know, but he wants to.

I leave home one morning and drive to central North Dakota, where I've heard a guy named Gabe Brown grows both cash grain crops and livestock using regenerative agriculture—a model that sounds very applicable to my family's ranch. The environment he's working in is similar to western South Dakota: little rainfall, formerly native prairie, cold winters and hot, hailstorm-ridden summers. I have to see for myself what he's up to, and not just for this book. I turn north on a gravel road that hops over Interstate 94 and settles into a rural landscape outside of Bismarck. Wild sunflowers line both sides of the road in thick stands several feet high. I drive with the windows down, and the smell of sap from a shelterbelt of tall pine trees fills the car. The day is cloudy and cool but still dry, a typical northern plains weather trick that makes farmers hope rain is on the way, but it never falls. When I spot the Brown's Ranch sign, I turn down a narrow gravel driveway.

I pull up to the yellow brick house and get out of the car, intending to knock. Instead I hear someone yell, "Stephanie!" and I turn to see a short, heavyset man smiling and waving from the nearby machine shed. He's standing in front of a stainless-steel chicken evisceration table, which thankfully has no eviscerated chickens on it at the moment. I walk over and we shake hands. "Just a minute, I gotta tell these guys what to do," he says, and then instructs two young men (I later learn they are interns) to replace a broken corner post in the pasture where the chickens are. After some small talk about my drive up and the weather—no midwestern conversation starts without this ritual—Gabe suggests that we jump in his Polaris Ranger utility vehicle and "go look at some soils." I would love to see some soils. I've been thinking and reading about soil for months in anticipation of this day. I've come to understand that soil fundamentally sustains me, sustains all of us. Soil is the key to human life. The nutrients we need to survive flow from the soil and into the food and livestock forage that grow in it. Healthy soil preserves water, resists erosion, and stores carbon (the basis of life on Earth and essential for all living organisms). Without healthy soil, we are nothing.

The link between soil, human health, and the environment is one of Gabe's primary concerns. Industrial agriculture makes that link weak, he believes. "I started studying a lot about human health," he tells me as we drive out to a field. "You look at the nutrient densities of the foods we produce. It's just plummeted. In my mind, that's directly related to the loss and degradation of our soil resources. Soils no longer have all the nutrients that they need." Why? I ask. "You talk to any grain farmer today, and they want to farm by a recipe card," Gabe explains. "I'll seed on this date, I'll seed a monoculture, it will need so many pounds of N, P, and K, then I'll spray this herbicide, I'll spray this fungicide on this date, and it's all off a recipe card. It's not that way. It shouldn't be that way. What needs to be done is you gotta look at each field and ask, what does that resource need? What is missing in that field? Then you start building back your soils, and as you do that the soil biology will make more nutrients available so the plants will be healthier, and the animals will be healthier, and then the humans that consume those products will be healthier."

This is Gabe's mission: to build back soils, or bring the levels of carbon, organic matter, and microbial soil life to what they once were. It's not enough to simply maintain soils because most are already degraded. Farmers need to go a step further by improving the soil, he says. "I hear it so often. People want to be sustainable, sustainable, sustainable. That's the cliché word; everybody wants to be sustainable," he says. "But why do you want to sustain a degraded resource? We need to be regenerative. If we are going to have healthy food and healthy soils for the next generation, and generations to follow, we got to build our soils back."

I feel sheepish and try not to let it show: I've said the same thing many times, that I want agriculture to be sustainable. But Gabe is right. Sustaining degraded land, water, air, and ecosystems through better agriculture is not enough. Agriculture needs to not only stop further damage, but also regenerate resources to preindustrial farming health or better. Regeneration comes from treating the farm and ranch as an interconnected environment, Gabe says. Diversity of life is necessary for a healthy ecosystem, and the same is true on a farm. "You gotta think of your operation, your ranch, as an ecosystem, because that's what it is," Gabe says. "It's the soil biology, it's the insects, it's the plants, it's the animals, all working together to be healthy, just like a native prairie ecosystem would have been three hundred years ago with the bison and everything."

Gabe reminds me of an old-time preacher, an association that isn't surprising when I learn that he spends the winter traveling around the United States preaching the gospel of regenerative agriculture to other farmers. People also trek to his farm every summer to see the restored soil and grassland, the free-range chickens and sheep, the grass-fed cattle, and the twenty-species cover-crop mixes. "It's to the point where we'll have two thousand people a summer through here," he says. "We've had visitors from all fifty states, sixteen foreign countries. It's gotten to the point where it's become our life. Like my phone, I'll average twenty to twenty-five calls a day. I'll spend two hours every night answering emails. During the winter, I'm pretty much gone from mid-October through mid-March; that's when I shut it off. I'm just traveling, speaking about soil health and sharing our story." And when Gabe gets to talking, there's no stopping him. When he's passionate or upset, his voice booms and his words rush out in a flood. His face reddens. When he's feeling sentimental or serious, his voice softens and his words slow. When he's indignant, he asks loud rhetorical questions that he answers. He's opinionated, but he listens and nods when others say their piece. He often speaks in jokes, sarcasm, and idioms. He uses colloquial language, and his speech patterns are distinctively midwestern. He challenges his neighbors about their conventional practices, and he's made enemies. He's also inspired people around the world to farm and ranch using regenerative principles.

From the pulpit of his Ranger, Gabe launches into the story of Brown's Ranch. He and his wife, Shelly, moved to the ranch—then owned by her parents—after college in 1983. They ran cattle and farmed cash grain crops conventionally, and they also worked in nearby Bismarck. Shelly had grown up on the place, and in 1991 she and Gabe bought the home section from her parents. Free to run the farm how they wanted, the Browns "went no-till" in 1993, meaning they stopped using plows to prepare the soil for planting or till harvested crops under. Instead, they leave the residue on the fields to break down and act as mulch.[1] "I was interested in saving moisture and time," Gabe says. By then they had 100 to 150 cow-calf pairs and were raising specialty breeding bulls. Wanting to try something different than the usual small grains like spring wheat, oats, and barley, Gabe planted some new crops in his no-till system. "I started planting peas, alfalfa, a few different crops," he says. "Then in '95 what happened, the day before we were going to combine—we had

1,200 acres of spring wheat in—we lost 100 percent of our crop to hail. Boy, after the hailstorm—I had no crop insurance, nothing, it was pretty devastating—I started thinking, what do I do for feed for livestock? So I started planting some sudangrass, millet, a few things, and of course couldn't pay back the operating note. So then the banker, he didn't want to loan me so much money the next year. Well, we made it through, thanks to off-farm jobs and that."

Since the Browns were farming conventionally at the time, the operating note Gabe couldn't pay included bills for fertilizer, agrochemicals, seeds, machinery, and fuel. Most conventional farmers take out loans every year and pay them back at harvest—*if* there is a harvest. When farmers specialize in one crop, like the Browns did in wheat, and suffer a 100 percent loss, then things can get ugly, especially without crop insurance. The Browns survived the year of loss, only to find themselves in another crisis the next season. "The next year then I started to diversify a little more, planting a few different species and that. Lost 100 percent of our crop to hail again. So that was two years in a row. Well then, things were really getting tough. We started to diversify even more, more alfalfa. At that time I didn't know what a cover crop was. We just were trying to grow feed for the livestock to stay alive. That was two years in a row. Ninety-seven comes along, and it was a drought. Nobody combined an acre.[2] So that was the third year in a row. Ninety-eight comes along—meanwhile I'm diversifying even more, I'm trying different species, I kind of got interested in growing different combinations of covers, two- and three-way mixes—'98 comes along, we lost 80 percent to hail. So we lost four crops in a row."

Desperate to save cash, Gabe had stopped buying fertilizer, pesticides, and herbicides. He was replenishing his soil and combating pests and plant diseases the only way he could afford to: naturally. Because the soil had been farmed conventionally for decades, the first few crops grown without inputs looked "just fair," as Gabe puts it. Slowly, though, the crops began looking better with every input-free year. Gabe also realized that the soil was changing. "We started to notice the soil health improving," he says. "I was pretty fortunate that when I first started we took some soil tests. These soils on this cropland here were from 1.7 to 1.9 percent organic matter.[3] I noticed that started to inch up. We'd also done some basic infiltration tests around that time, and we could only infiltrate a half of an inch of rainfall per hour. That was pretty poor. So if it rained over half an inch, it was just running off. From all the years of tillage, that's what happened. Well, we noticed the more we grew these covers and the more diversity we had, that the soil was infiltrating better and it was also cycling nutrients."

More organic matter, better water infiltration, and improved nutrient cycling: the silver lining of what many people would see as a dark time. Those years allowed Brown's Ranch to emerge stronger and more self-sufficient. Not that they weren't hard—Gabe jokes that they were so broke the banker knew when they bought toilet paper—but they led him to realize something that fundamentally changed the way he operated: *he did not need inputs.*

20

The Soil

Gabe didn't transition all at once, nor did he do it alone. He sought help from the local Natural Resources Conservation Service (NRCS) office and other farmers and ranchers using regenerative models. "It's been a long learning process," Gabe says. "I was really fortunate that I met the right people at the right time." One of those right people was Dr. Kristine Nichols, who at the time worked for the USDA Agricultural Research Service (ARS) Northern Great Plains Research Laboratory across the Missouri River in Mandan. Nichols has spent much of her career studying soil biology, particularly the union between plant roots and fungi. From the days of Earth's first photosynthetic organisms, plant roots and fungi have enjoyed a symbiotic relationship, an association called mycorrhizae (from the Greek *mykos*, or fungus, and *rhiza*, or root). Mycorrhizal fungi live in and extend from root tissue, bringing nutrients and water to the host plant, suppressing weed growth near it, binding the nearby soil into aggregates that hold water, and probably dozens of other activities that we don't fully understand, since scientists haven't been able to culture even the most common mycorrhizal fungi in labs for further study. The fungi simply die if they aren't attached to plants. Plants return the fungi's allegiance by supplying energy from photosynthesis and, some scientists venture, important growth hormones.[1]

Some mycorrhizal fungi form microscopic tree-shaped structures called *arbuscles* within their host's root tissue. These arbuscles shuttle nutrients from the fungi to the plant. Fungi that create arbuscles are called arbuscular mycorrhizal (AM) fungi, and they exist on most temperate and tropical plants and crops virtually everywhere in the world except a few arctic areas.[2] AM fungi perform their plant-assisting functions in part by producing a substance called glomalin. Nichols is a leading researcher in glomalin and how crop rotation, tillage practices, organic production, cover crops, and livestock grazing affect the way AM fungi produce it.

Aside from the relationship between plant roots and fungi, researchers also look at the integrated, three-tiered system of "consumers" chomping away at organic material, turning it into nutrients. First-level consumers, like microbes, the smallest and most numerous soil citizens, turn big pieces of organic matter into smaller ones. Secondary consumers, such as protozoa, eat these first-level consumers or their waste. Third-level consumers, like ants or beetles, in turn eat the secondary consumers. All this eating creates soil that contains abundant nutrients easily accessible to plants.[3]

If soil biology seems daunting, don't worry—it is. Soil is one of the most complex substances on Earth, if not *the* most complex. Humans understand only a tiny fraction of what's actually going on underground, but we do know that billions of microorganisms work together to make soil a living substance. Ecologist David Wolfe explains it this way:

> Step out into the backyard, for example, push your thumb and index finger into the root zone of a patch of grass, and bring up a pinch of earth. You will likely be holding close to one billion individual living organisms, perhaps ten thousand distinct species of microbes, most of them not yet named, catalogued, or understood. Interwoven with the thousands of wispy root hairs of the grass would be coils of microscopic, gossamer-like threads of fungal hyphae, the total length of which would best be measured in miles, not inches. That's in just a pinch of earth. In a handful of typical healthy soil there are more creatures than there are humans on the entire planet, and hundreds of miles of fungal threads.[4]

The thought of a handful of soil containing so many life forms—more than there are people on the Earth—is mind-boggling, as are the gaps in our knowledge of soil. As a layperson, I have only a rudimentary understanding of soil biology, and it's safe to say most people are like me. The average person doesn't need to be a soil scientist, however, to recognize that invisible forces make plant life possible, which in turn make animal and human life possible. It also doesn't take a scientist to see that human life depends greatly on the billion organisms in that pinch of soil.

Still, it's good that some people understand more of what's going on and can articulate its importance, like Gabe's friend Kristine Nichols. Nichols, now with the Rodale Institute, spent a lot of time on Gabe's farm in the early 2000s, observing the effects of his no-till farming, cover crops, and reduced input use. Back then he was still using small applications of synthetic fertilizer and herbicide, but he hadn't used pesticides or fungicides since the 1990s. Gabe said Nichols convinced him to quit using synthetic fertilizer once and for all as her research revealed its negative impact on glomalin production in AM fungi.

"She said, 'Gabe, you've come a long way, but your soils will never be truly sustainable until you remove the synthetic fertilizer. Because with the synthetic fertilizer, your soil biology populations won't propagate to the numbers needed to convert organic material into inorganic plant-usable forms.' So from 2003 to 2008 we did split trials where we would fertilize half a field, not fertilize the other half. In four years, the unfertilized yields were equal or greater than the fertilized. There's been no synthetic fertilizer here since 2008. We're not organic; we still occasionally use some herbicide.[5] But I'm down to about one herbicide pass every two to three years. With our yields, our proven dryland corn yield is 127.[6] The county average is 100. So we're about 25 percent greater than county average without all the inputs."

"The thing of it is, though, you can't do that with just a corn and bean rotation," Gabe continues. By "that" he means produce yields that are 25 percent higher than county average by growing just two crops over and over. It takes diversity to build soils that are rich enough to produce high yields. "We grow—oh man, if I listed all the crops—oats, barley, sunflowers, corn, alfalfa, winter triticale, hairy vetch, all of our cover crops, and there are a lot of others, rye, I'm not listing even all of them. Peas. We don't grow all of those each year, but the same field won't have the same crop type for a number of years. We are rotating all the time."

In addition to yearly rotations, he also does seasonal rotations, so that each field is home to multiple crops within a year's time. "We try on every field, every year, to grow a cover crop besides the cash crop," he continues. "That might be before a cash crop, after a cash crop, or along with a cash crop." *With* a cash crop, as in two crops growing at the same time? How does that work on one piece of land? I ask. "I seed the cover crop along with the cash crop," Gabe explains. "For example, I seed clovers with an oats crop. The oats is taller than the clovers so I can straight combine the oats, leaving the shorter clovers to continue growing. Corn with hairy vetch would be another example."

Gabe's fields are also smaller than those on most conventional farms, which creates more variety on the land—think a patchwork quilt instead of a solid-color blanket.[7] These small fields plus his combination of intercropping and multicropping better reflect the native prairie and are the best way to regenerate soil because there's continuous life in the fields. "That's the key to healthy soil, is something living all the time," Gabe says. "Look at your native prairie soils; there's something living all the time, obviously until the snow comes, and even then they're still alive. That's the problem with farming today: people are in monocultures and you don't find that in nature. Where is there a monoculture? There isn't one. It's very seldom you find a monoculture."

Something living all the time—a simple rule, but a hard one to follow under the conventional model. Gabe wouldn't call a monoculture "living" in the sense that his diverse fields and pastures are alive (teeming is more accurate) with insects, earthworms, birds, soil microbes, and more. The agrochemicals required to grow monocultures kill good as well as harmful insects, most of the soil biology, and all plants except the desired crop. The chemicals essentially sterilize the field, meaning the members of the three-tiered soil system practically disappear because there's little to eat. The only thing left is the corn, wheat, soybeans, tomatoes, or whatever the specialized crop. The monoculture itself is alive, but there's very little life happening among it—the ecosystem is gone, both above ground and under. That's why the monoculture model, even on "sustainable" farms, is so harmful.

The monoculture model kills the soil in other ways, too. On such farms the land lies fallow, or unplanted, when the cash crop isn't in the ground. Even weeds don't grow because farmers use herbicides to "burn down" the fields. Sometimes fields lie empty all growing season in a practice known as summer fallow. Farmers leave the field unplanted and either spray or till to control weeds, with the good intention of letting the land rest. Just a few days before visiting Gabe, I rode my horse in the summer fallow field behind my parents' house, just for fun. Riding over the brown, lifeless field, I remembered that barren is the ideal look for summer fallow. People talk scornfully about farmers who let "trash" (weeds) grow in their fallow; they see these farmers as lazy. A bare field is a clean field and thus a desirable one, the thinking goes. Gabe explains why this practice isn't resting soil, but taxing it: "There are more microorganisms in a teaspoonful of soil than there are people in this world. Think of that," he says. Gabe echoes Wolfe whether he realizes it or not. "So you've got a little bit the size of your thumb, and there are more organisms than there are people in the world. What are they going to eat? They have to eat from a living plant root. If you summer fallow, you don't have nothing alive. There's nothing to feed them."

One problem with summer fallow is underground—the soil's microorganisms starve—but another problem is above ground, and that problem is soil exposure. Soil erodes when farmers leave it bare: wind carries it off and water washes it away. The soil also dries out. "People think they're saving moisture by summer fallowing," Gabe says. "You're not, because what stores moisture is that organic matter because that soaks water up like a sponge and it's there. Well, the only way to grow organic matter is by growing something. So you're actually better off to grow a cover crop in those off years, and if you have livestock then you convert it to dollars by grazing it, and you will store more moisture long-term than you will in summer fallow."

Cover crops include plants like turnips, peas, sorghum, clover, and countless others. Anything can be a cover crop if it helps the soil in some way, even plants that conventional farmers call weeds. I read about a farmer who planted yellow mustard, which people typically call a weed, as a cover crop because disking it into the soil replenished sulfur levels.[8] Good luck finding a cover crop in the corn-heavy Midwest, though. Despite the benefits cover crops provide, conventional farmers tend to see them as a waste of time and resources because they do not generate calculable farm income; they usually can't be harvested and sold the way cash crops are. Some covers can be hayed, but most farmers sold their haying equipment long ago, when they got rid of their livestock. Some can be combined for seed, but farmers usually lack these specialized machines, too. In other words, farmers can't transform most cover crops into commodities. Cover crops do not fit into the industrial model because farmers can't quantify their benefits on a balance sheet. From the industrial viewpoint, spending money and time planting a crop that doesn't turn a profit is unwise (with the exception of crops that are subsidized by the government, as corn and soy are). The industrial model only compares inputs versus outputs and doesn't consider intangible gains like soil health, water infiltration, nutrient cycling, or the life of billions of microorganisms.[9]

For the regenerative farmer, though, planting cover crops for the intangible benefits is worth it. On a diversified farm like Gabe's that also includes livestock, cover crops are doubly worth the effort because they *do* have a tangible value: they can be "harvested" by livestock. When livestock graze a cover crop, they absorb the nutrients and convert them into meat, milk, or eggs—commodities a farmer can sell, which means they "market" the cover crops through the livestock. Gabe tells me he can also market a failed cash crop through the livestock. "People say, 'Gabe, how can you farm with no crop insurance?' Livestock are my crop insurance. If I get hailed out, the livestock will move on those acres and we'll convert it to meat dollars." Livestock aren't the only method for marketing cover crops: Gabe harvests the seed from some of his covers and sells it to other farmers because such mixes are practically impossible to buy. In fact, Gabe has to purchase most seed varieties individually and mix them himself in a giant seed mixer.

But why can't we create a more direct market for cover crops that doesn't always require livestock to ingest them first? asks Dan Barber in *The Third Plate*. Barber envisions a new American cuisine built around everything the land provides, including cover crops. Cover crops such as millet, flax, buckwheat, and peas—plants that make the soil fertile enough to support our favorite items like wheat, fruits, vegetables, and meat—should be integrated into our diet, Barber argues. Livestock that help with soil fertility and other farm functions should be included as well—not just the choice cuts, but the entire animal. He calls on chefs and consumers to cook with the whole farm instead of just with prized or familiar items, such as steak, winter tomatoes, or enriched flour. Cooking with the whole farm will create a market for covers and encourage farmers to grow them, which will financially benefit farmers, increase soil health, and create a national cuisine that rewards regenerative agriculture instead of the production of a small number of cherry-picked crops and meats.[10]

Gabe and I stop at a field with a cover crop on it, a mix of twenty-one species. Some I know—clover, alfalfa, and sudangrass—but most I've never seen before. Gabe walks into the field and shows me different plants. "This is sorghum. The sudan you can recognize. Here's buckwheat. This here is a common vetch. There are clovers growing," he says. He points out a type of edible kale, then a grass called teff. "Now think, Stephanie, if I was to grow a monoculture," he says. "Say this was just sudangrass.

That biology would only be eating root exudates from sudangrass.[11] Now with twenty-one different species, think of the diet that biology has. They are going to propagate numbers and they're going to be much healthier."

He looks out over the field, which is thick and green. Unlike a monoculture corn or wheat field, the plants vary in height and appearance. Instead of a flat, even expanse similar to a mowed lawn, the cover-crop field is an undulating mosaic. It looks almost like native prairie. "This doesn't look bad," Gabe says. That's the modest midwestern way of saying it looks amazing. "This was just seeded here about two and a half weeks ago. It's had no fertilizer, no chemicals, it just is what it is." Gabe's cover crops are essentially giant salads for the cattle. These cow salads are high in nutrition because of the plant diversity, which helps the cattle put on weight and fend off sickness, the same benefits Phil's buffalo enjoy when grazing diverse native prairie. In turn, the cow manure and urine mineralize the soil, naturally replenishing the nutrients removed during grazing.

Cover crops also provide homes for beneficial insects. "This one here is fusilli," Gabe says. "This plant here, that'll produce a big purple flower. Cattle don't eat this, but it's in here for that flower because we want to attract the pollinators and the beneficial insects.[12] We try and have, in all of our mixes, flowering species to attract pollinators and the beneficial insects. That's all part of an ecosystem. Look at the prairie. If you've got true native range, you always have some flowering species. That's what we're trying to do. We're just trying to mimic nature, is all we're trying to do here." Beneficial insects can be pollinators, but they can also be predators such as wasps and ladybugs. Predators find cover crops inviting because they contain lots of bug prey. These predators help control insects that harm the cover crop, making pesticides unnecessary. Even in Gabe's cash crops, like corn or wheat, predator insects are still present, attracted by nearby cover fields, native prairie, or other crops he has seeded among the corn or wheat. These predator insects do a better job than pesticides of controlling bugs that snack on the cash crops, Gabe says. "People can ask me, 'Gabe, how can you get by without spraying for insects?' Well, it's because I got the home for all the predator insects," he says. "You look at our plants, yes, you can find a few holes in some leaves and that, but it's never to the point that it's economically detrimental to us because we have the lady bugs and all the good insects that take care of that."

In nature, he says, beneficial or neutral insects far exceed the number of pests (which are only defined as such because they harm human endeavors, not necessarily the environment). Gabe cites work by Jonathan Lundgren, a research entomologist at the USDA's Agricultural Research Service facility in Brookings, South Dakota. According to Lundgren, Gabe tells me, for every insect that's considered a pest, there are 1,700 other species that are either beneficial or neutral. "Think of that!" he exclaims. "My neighbors who farm forty thousand acres, the airplane's flying over all the time spraying something. Fungicides, pesticides. He's targeting that one pest, but he's killing all 1,700 other ones. I'm not saying they all live right in one place, you know, but you know what I mean."

When I look into Lundgren's work later, I find out Gabe is way off—for every one species considered a pest, there are perhaps as many as 6,000 species that either help people or contribute positively to the ecosystem.[13] In many cases, we don't understand what exactly they do to help us, but we know they aren't hurting us. By contrast, between 1,000 and 3,500 insect species are labeled pests. Gabe is right in saying that not all beneficial or neutral bugs are in once place at one time, so it's not like a pesticide pass kills thousands of species. But it does kill hundreds and, depending on the chemical, prevents others from coming back. As Lundgren writes, "Pest management shouldn't throw the baby out with the bath water; controlling a few pests should not come at the expense of beneficial species simply because we don't understand the benefits they provide."[14] Because pesticides kill every insect in their wake, they fumigate a field in the same way an herbicide does: with total destruction.

The absence of bug life eventually causes environmental imbalances. Without predators and beneficial insects, pest populations explode out of control, something nature would rarely allow since its model of checks and balances does a good job of keeping bugs in line. Monocultures also contribute to bug population imbalances because they encourage the presence of the very bugs that love snacking on them. Corn leaf aphids prefer corn, for example. When farmers plant a huge field of corn, leaf aphids can show up en masse; it's what they evolved to do. When practically every field for hundreds of miles is also planted to corn, one can see how easily the leaf aphid population can get out of hand.

Now that some insects are immune to pesticides, relying on chemicals to control them makes even less sense. Worldwide, more than five hundred insects and arthropods are resistant to agrochemicals.[15] But conventional farmers simply apply stronger concentrations, mix different chemicals together, or turn to older, more toxic chemicals. It's what they were taught to do by their parents and grandparents. Pesticides ensure minimal insect damage, which in turn ensures more profit, the thinking goes. Most farmers have zero tolerance for pests and even less of a desire to gamble, as they see it, by not applying pesticides. They're right in that sense: growing in monocultures without pesticides is almost impossible. It will take a drastic change in perspective for farmers (and ranchers, for that matter, who spray hay fields and pastures for insects) to trust that nature can control pests better than companies like Syngenta, Bayer, and Monsanto. It will take a similarly drastic change to stop the monoculture system that makes chemicals necessary in the first place.

Gabe is convinced that most conventional farmers feel bad about using agrochemicals, as much as they defend their need to do so. "Does your dad have row crops and that?" Gabe asks me. I say yes, that he's fully invested in the industrial model. Gabe nods, not condescendingly but empathetically. "If you ask him, 'Dad, how do you feel when you go out and spray something?' I'd be willing to bet, deep down, it bothers him. It will everybody. If that will kill life, then why do we want to feed it to people? Now, I don't want to be a hypocrite because I told you that I still use herbicide, too, now and again, so I'm not trying to tell you I'm above any of that. But it bothers me every time we do it. You look at a lot of the farmers today and they do a pre-plant herbicide, they do an in-crop, they do a burn down postharvest and it's like, 'My goodness, that can't be healthy. It's got to be getting into the food.' I just think we got to get away from all of it," Gabe says. He pauses. "I'll be honest with you; there are a lot of times I'm tempted to do a little tillage so I don't have to do that herbicide pass. Then I get to looking at how I've gotten these soils to such a point that I don't want to go backwards. I wrestle with that all the time. It's tough."

Gabe is getting to a question I've been meaning to ask: why hasn't he gone USDA certified organic? The answer is in the soil. One way organic farmers control weeds is through tillage—but Gabe is firmly no-till. Instead he relies on crop rotations and plant residue that stays on top of the soil to prevent weed growth. It isn't easy to keep a thick, consistent layer of residue, however. "It's not all roses here," he tells me. "We battle some perennial noxious weeds because in our system I'm only using an herbicide every two to three years. We have sixty earthworms per square foot. You get soil alive like that, and you're cycling nutrients so fast that I can't keep enough of that residue on the surface. Then we get some weeds."

Gabe keeps the herbicide option in his toolbox for extreme weed issues, and it's the one practice stopping his operation from being certified organic. Even a single application every two or three years breaks the rules. While Gabe's pastures are organic, his fields aren't, and because the cattle, sheep, and chickens eat the cover crops, they aren't technically organic, either. I think of Kevin's commitment to the organic standards and his doggedness in finding ways around issues like weeds and disease. I ask Gabe directly: couldn't you find organic ways to overcome the noxious weeds? Couldn't you earn USDA organic certification if you *really* wanted to?

He's silent for a moment, then says, "I haven't been smart enough to figure out how—and I'm serious about this and I've been trying—how do I do away with the herbicide so I can be organic, but not till. Because I will . . . not . . . till." He pauses for dramatic effect between each word.

"The other thing, though, even if I could, I probably would not because I don't believe in somebody else putting a stamp on me." Like Phil, Gabe has a wide independent streak—most farmers and ranchers do in some way—and that makes him uncomfortable with submitting to the authority of a certification organization. Except for the herbicide, he operates organically by choice, so he doesn't see the point in obtaining certification. To him, the proof is in the products he sells at the farmer's market, not in a label. "My customers know," he says. "We invite them all, we say, 'You come out, anytime you want, to see how those chickens are raised, you want to see how the beef is raised, you come out and let's do it face-to-face.' I don't need a third party putting a stamp on my operation. I want to build enough rapport and trust with my customers that they know that's what it is."

To label or not to label—that is the question. As an urban consumer who buys roughly 90 percent of her food at grocery stores, I want a label. I don't have a farmer telling me about the product, and I don't trust corporate food conglomerates who might try to convince me a product is organic without a USDA label. As an advocate for farmers and ranchers, though, I want producers to operate according to their principles because they are the ones on the land. They are responsible for what is gained or lost. Obviously they don't operate in a bubble; corporate pressure, legislation, and consumers partly dictate their actions. But whether nonfarmers like it or not, farmers and ranchers are free to choose how they want to operate for the most part, even if that means continuing to use conventional practices. Nonfarmers can encourage them to operate differently or, as I hope young people will increasingly do, they can join the system and become farmers who do things regeneratively—change from within.

It's not terribly productive to bicker over whether to certify farms as organic or not. What would be more productive is action. In the end, I hope labels become unnecessary due to a nationwide shift away from industrial agriculture and toward regenerative agriculture. Such a shift won't happen unless we commit to it in the same way we committed to industrial agriculture: wholeheartedly and with an unwavering focus that isn't on yields this time, but on growing the most wholesome food possible. We need widespread change, a turn of the tables so that most of our nation's farmland and grassland is organic or practically organic and conventional land is the exception. Most of all, we need this change to stem from the inside, from the hearts and minds of growers who believe or have come to believe, like Gabe does, that mimicking nature is better than fighting it.

21

The Abundance of an Acre

Gabe and I pull up to a field with an oat-pea cover-crop mix that was "mob grazed" by cattle. *Mob grazing* is an intensified version of Phil's holistic management system. Under mob grazing, ranchers move livestock at least daily, usually more often, between small paddocks split with portable electric fencing. Animals either eat or trample every plant, and usually they don't return to the same paddock for a year or so. It's rotational grazing on steroids. To most observers (including me until I started writing this book), a mob-grazed field looks like a disaster. But that's the point. Now that I am familiar with holistic management, I notice hoof prints and cow patties spread evenly over the field we're touring. Gabe says crushed forage and evenly distributed animal impact are exactly what he's looking for when the cattle leave a paddock. Even though the plants are trampled, he says, they still cover the soil so it doesn't dry out, plus the roots hold it in place against the strong North Dakota wind. He says within a week or so he'll seed the field to something else, right into the trampled plants.

Mob grazing was the final component of Gabe's regenerative model for cropland. Removing synthetic fertilizer, tillage, and agrochemicals and adding cover crops and diverse rotations helped, but weren't enough. Animal impact, Gabe realized, was the difference between great soil health and excellent soil health. For farmers practicing no-till, livestock are especially important because they provide soil ventilation in place of a plow. And with the addition of livestock, the farm becomes a true ecosystem, with soil biology, insects, plants, and animals working together. It's not just the 350 cow-calf pairs and 400 to 800 yearlings that participate in the ecosystem—the Browns run other species, too. "My son came back from college five years ago, and even before he finished college he said, 'You know, Dad, we got all this diversity of plant species, but we just have beef cattle. We need diversity in livestock, too.' He wanted to start laying hens, so he started a laying hen operation," Gabe says. "Now we've added sheep, we've added pigs, broilers, ducks, and turkeys. We don't have ducks and turkeys this year. All we're trying to do is, well, think of what you had on the native prairies. You had bison and deer and elk and you had all the birds following them. You had all these different species."

Having multiple enterprises on the same land is known as *stacking*, a process Michael Pollan defines as "mimicking relationships found in nature and layering one farm enterprise over another on the same base of land . . . farming in time as well as in space—in four dimensions rather than three."[1] Farms that stack have financial resiliency; if one enterprise fails, the farmer still has income from other enterprises. Most farms only have one or maybe two enterprises, which is the equivalent of placing all of one's eggs in a single basket. Gabe saw this firsthand when his wheat crop failed four years in a row and he had nothing else to sell.

The worst part about the limited-enterprise model, though, is that it isn't an efficient way to feed people. Gabe explains it this way: "I get really frustrated about how one of the buzzword phrases is, 'How do we feed nine billion people by the year 2050?' That's absolutely no problem. It's no problem if we stack enterprises. The problem comes in that, with the current production model, we're only producing one commodity, only so many kilocalories of energy, off of an acre, whereas if you stack enterprises like I do, you're going to be producing many, many more nutrients on a per acre basis. So feeding the world is absolutely no problem if you change production models. It is a problem if you're using the current production model because it takes so much fossil fuel that it's not regenerative, it's not sustainable, and it makes no sense." So an acre of land on a regenerative farm utilizing stacking produces more food per acre with fewer input costs than on a conventional farm. This is true efficiency, the real answer to feeding a growing population.

Gabe offers an example of the danger that the conventional, limited-enterprise model holds for the future. He recalls a young person from northeastern Colorado who approached him after a speaking event. The young man asked Gabe how he could convince his dad and grandpa to diversify and be more holistic, as Gabe had urged during his talk. Gabe asked what the farm's crop rotation looked like. The young man said the family had been on the farm since the 1920s and they had never grown a crop other than wheat. Gabe said he imagined the soil must be dead. The young man said yes, it was, that they couldn't even get twenty-bushel wheat anymore.[2] Gabe asked how they were surviving. The young man said everyone had off-farm jobs. "This is what's happening all over," Gabe says. "That's why with production agriculture in the Midwest, they have so many inputs, inputs, inputs because it's reduced to monoculture. That's all it is. We're reduced to this mentality that that's all we can grow and that's all we can do."

In other words, farmers are convinced that they can only grow one or two things on the land, and these only with help from expensive inputs. In reality, the land could yield far more if farmers returned to diversity. Stacking enterprises has also allowed Gabe to reduce the farm's size, meaning he's more productive on fewer acres (and less stressed out by the work load). He used to operate 6,000 acres; now he's at 5,000, and he continues to downsize. Of those acres, he owns 1,400 and leases the rest. He plans to let go of some leased land in the near future. "We've got way too many acres," he says. "Paul and I honestly believe that we could easily support a family on a quarter of land, easily, with all our enterprises.[3] We have so many different things going on."

Adopting regenerative agriculture nationwide would allow more families to live on the land. Farm consolidation under the industrial paradigm displaced hundreds of thousands of rural families, people who got out instead of getting big. Gabe points to the example of his neighbors, three brothers who together operate around 40,000 acres. If Gabe's claim is correct that 160 acres could support one family—when stacked with enterprises and managed regeneratively—then those 40,000 acres could potentially sustain 250 families instead of three. Even if we bumped the farm size to 1,000 acres per family, just under the North Dakota average farm size of 1,268 acres, an impressive forty families could make a living there.[4] When we look at examples like this, it's not hard to understand why our rural communities feel hollowed out. There are hardly any families left on the land anymore. No families actually live on Roth Farms, for instance, or on the sprawling, corporate sugarcane operations that cover the Everglades Agricultural Area. If we stacked enterprises and farmed more productively, then we could revitalize rural communities and take pressure off stressed urban areas.

If nothing else, being more productive on fewer acres with limited expenses makes farming enjoyable, something conventional agriculture hasn't been for decades. Gabe doesn't worry about fluctuating input costs, volatile commodity markets, or hefty loan payments. His primary concern is the soil. If he takes care of that, the rest of the farm will take care of itself for the most part. I remember the financial pressure Ryan Roth faces and the stress he feels, the razor-thin profit margins and the prospect of losing the farm to subsidence. I'd take Gabe's job over Ryan's any day. "It's a lot of fun," Gabe says. "It's gotten to the point now that—how do I say this—I don't mean to sound arrogant or that, but it's really difficult for us not to make money. One of our mottos is that we like signing the back of the check and not the front. We just don't have all those expenses." Let's be honest: making money is a whole lot more enjoyable than not making money. Why wouldn't farmers want to reap these financial rewards and work smarter, not harder?

Selling conventional farmers on the concept of stacking, however, is a little like attempting to convert someone to a new religion. Most farmers have internalized the first commandment of industrial farming:

specialization. Going back to the "old days" of cover crops and livestock is considered a backward move. When Gabe speaks in the Corn Belt about adding livestock, farmers think he's crazy. At an event in Indiana, where everyone grew corn and soybeans only, the crowd overtly laughed when he discussed running cattle on fields. "I said, how much does the average producer spend in fertilizer every year?" Gabe says, describing his response to their laughter. "Well, if you get your soils healthy and integrate some livestock, you wouldn't need that. I'm not saying you could eliminate all the fertilizer with just that, but a good percentage of it. What's that worth? Then I ask, doesn't anybody in Indiana eat beef? Why not get some on your farm?"

If saving money doesn't persuade the crowd, Gabe tries another tactic: appealing to the farm's future. Most people dream of passing the farm on to their children, but conventional farming parents are finding that the farm barely supports them. As is, it won't support sons or daughters, their spouses, and eventually grandchildren. That should never be the case, Gabe says, especially on today's farms that span thousands of acres. When farmers limit what their land can produce through specialization, they limit future generations. "A pet peeve of mine is when I hear, 'We can't have Junior come back to the farm because there's no room,'" Gabe continues. By "no room," parents mean there's not enough land for everyone to make money. "I say, do you have livestock? Perfect opportunity for a young person. Any type of livestock. People, they put these blinders on."

Blinders are a great metaphor for the conventional agriculture mentality. For those who did not grow up reading just about every horse-related novel out there like I did, racehorses and horses that pull carriages wear blinders, which are leather or plastic cups attached to the bridle that prevent the horse from seeing to its rear and side. The horse can only see what's directly in front of it, which prevents it from being scared of or distracted by crowds or other horses. The blinders of conventional agriculture do the same by shutting out the past and narrowing the farmer's focus to one or two crops and their yields per acre placed against input costs. The farmer can no longer see things like soil health, the environment, consumer health, and water quality—these are the distractions conventional agriculture seeks to remove from the farmer's vision. It's like the young man's family from Colorado: the blinders of conventional agriculture prevented them from seeing that they could grow something other than wheat. These blinders make diversification—not only with livestock, but also with other crops—an invisible concept.

Gabe's method of running livestock on his cropland is at once ordinary—humans have done it since ancient times—and novel, since the idea of putting livestock on cropland is unorthodox to longtime conventional farmers and the new generation alike. Of course, the potential level of livestock integration depends on a farm's environment. Farmers can identify where their environment falls on Savory's nonbrittle to brittle continuum and cycle livestock accordingly. In his chapter on cropping in *Holistic Management*, Savory discusses the role of livestock on farmland, particularly in relation to after-harvest residues. He points out that "plowing raw organic matter into the soil does more damage than good. . . . Animals, on the other hand, will reduce the residues to dung and urine and still leave a mulch to cover the soil. Poultry also consume insects and help to keep their numbers in check."[5] He warns producers not to allow livestock to stay in a field too long because they will eat too much residue (leaving none for mulch) and over-churn the soil (because farmed soil is looser than grassland sod). Concentrating animals on small crop areas for very short periods works best, he writes. That's exactly what Gabe does with mob grazing.

It's not impossible for farmers to build back soils without livestock. Farmers like Klaas Martens, featured in Dan Barber's *The Third Plate*, used cover crops and rotations for years to regenerate soil, though he eventually added animals. Kevin's use of compost is another example. But spreading compost is not a realistic option for farms that span thousands of acres. Plus, a cover-crop/rotation system without livestock usually doesn't replicate nature's way of building soil. Animals are part of the soil system in most environments. Livestock make the farm a real ecosystem, something it hasn't been for many decades.

But to me, the most exciting part of livestock integration on cropland isn't the soil benefits (sorry, soil enthusiasts) or the economic gains (sorry, bookkeepers).[6] It's the opportunity such a model provides for dismantling the CAFO system. If farmers welcomed livestock back to the farm, then we could stop confining cattle, hogs, and poultry; these animals would reach slaughter weight on the land instead. Livestock raised on midsize or small regenerative operations would be grass-fed instead of corn-fed, and producers could tailor animal species to their specific environment for maximum benefit, creating more variety in a market dominated by beef, chicken, and pork. The destructive ecological and social consequences of CAFOs would start to disappear, livestock would live more dignified lives, and consumers could eat meat knowing that livestock help the environment instead of harm it. Putting livestock back on America's cropland could finally shut down the CAFO.

So are mob grazing, cover crops, and diverse rotations working? It's time to look at some soils, this time more closely. We get out of the Ranger and Gabe grabs a shovel. He jams the shovelhead into the ground and bends down to examine what he's unearthed, fingering through the soil for earthworms. "Last year, we did earthworm counts and we were averaging sixty per square foot. That's a lot of earthworms," he says. "When we started there was zero. You'd never find an earthworm because of all the tillage." We don't find any earthworms today, but there's plenty more to see in this soil. Gabe reminds me that the organic matter composition of the farm's soil used to be 1.7 percent to 1.9 percent, very low. "This field last July was 6.1 percent," he says, beaming with pride. "We've tripled the organic matter levels, which is carbon stored in the soil. Carbon is what all that soil life eats in order to convert it to forms that we can use." He grabs a handful of soil and holds it out for me to inspect. "That's what you want soil to look like."

The soil is black like the muck in Belle Glade and damp in my hands, even though Gabe says the ranch hasn't received any rain lately. It's not loose and fluffy, but clumpy, caked together in small chunks that break off against my fingers. "If you can imagine, this soil was just dull and gray and lifeless from all the tillage," Gabe says. "Now it's like cottage cheese. See all the soil particles, how it looks like black cottage cheese? That's what you want. So now you think, if a raindrop falls, it's going to go right through there. Where we could only infiltrate half of an inch per hour before, now we can infiltrate eight inches an hour. Which is unreal."

Eight-inch rains are rare in North Dakota, but that is precisely why it's so important for the soil to be absorbent. When a heavy rain hits, then the soil needs to be like a sponge so the water doesn't erode the ground or cause excessive flooding. Gabe mentions a section of the Red River, around Fargo, with a history of floods, some catastrophic to the city. Every spring, residents await the rising water. Gabe sees the Red River flooding not as a problem with the river or with excess rain or snowmelt, but with the soil in the Red River valley. It used to have about 8 percent organic matter; now it has less than 2 percent because of industrial farming. Gabe believes the Red River has flooded more often in recent decades because the soil in the valley and other drainage areas don't contain enough organic matter to absorb water, so it runs off. "We waste money on disaster relief for floods and then for droughts when we wouldn't have to if the soils could just absorb the rainfall to compensate for high levels of rain and store water for when there's little rain," he says.

The Red River is not a perfect example. The river floods for other reasons, such as the fact that it disregards the Continental Divide and runs northward. As springtime temperatures rise, the river carries snowmelt and ice north, right into areas that haven't thawed yet, instead of carrying water south and away like other rivers. When the snow and ice finally start melting in the river's northern sections, usually near Fargo or Grand Forks, the river already contains so much backlogged water and ice that it floods. Another problem: the Red River valley is a valley in name only.

It's pool-table flat. When the Red River breaches its banks, the water fans out every which way.[7] These anomalies contribute to the river's floods, but the soil certainly plays a role, too. The valley is divided into farms as far as the eye can see, and these soils can't absorb near the water they once did. More than 95 percent of the valley's native prairie, once the sponge for excess water, is gone.[8] Increasing water infiltration in the Red River valley wouldn't solve the flood problem, but it would no doubt help. I wonder how many other floods we could prevent or reduce by regenerating soil.

With the exception of those living along the Red River, though, flooding isn't the North Dakota farmer's main concern. Dry years are more common than wet years in the Dakotas, and even in an average year rain might not fall for weeks or months. The problem is that soil in conventionally managed fields neither absorbs water when it falls, nor stores it when it doesn't. Because of synthetic fertilizer, monocultures, agrochemicals, and tillage, it lacks the organic matter that creates soil particles, the clumps I see in Gabe's soil. Rainwater washes away, and when droughts come, the soil has no water stored for the crops. Unfortunately I saw this firsthand; my high school years coincided with an intense four-year drought in western South Dakota. No one had a crop. Our wheat was so poor that many fields weren't worth harvesting for grain—Dad had me swath them for hay instead. In a cab-less swather, I crawled across the dusty fields, a rag over my face and a dirt cloud trailing behind. I had to set the swather's header so low to the ground—the wheat was incredibly short—that I'd often doze it into the dirt and get stuck. I carried a fire extinguisher at all times in case I hit a rock and sparked a fire. Our pastures wore down to practically nothing (everybody's did). We sold some cows. We ran short on hay. I remember fields blowing, the wind carrying off the topsoil.

In dry years like those, Gabe's soil reveals its resiliency. His crops and pasture thrive while his neighbors' land withers in the heat. One of those dry years occurred last summer. Near the end of the summer, Gabe hosted a tour group of four hundred people, one of the largest tours he's ever done. From June 1 until August 20, the day of the tour, the ranch had

had 0.38 inches of rain—practically no moisture. "Yet we stood down in a cover-crop field that was waist high," Gabe tells me. "They said, 'How can you grow that?' I said, 'It's not how much rain you get; it's how much infiltrates and then how does it move through the soil.'" I think of what Phil said when he showed me the picture of the water runoff from his neighbor's pasture. *Is it a problem of not getting rain, or not using the rain that we get?* By now it's clear to me that as a nation we're not using the rain we get because we've degraded the soil. Gabe's black cottage-cheese soil drank up every drop of that summer's meager rain, plus it contained reserves. That wasn't all, though. The cover-crop field was not a monoculture—far from it. Fields with multiple species are more resilient during droughts because the plants can access moisture at different levels, Gabe says. "There are all different root types," he says. "There are taproots, fibrous roots, shallow, medium, deep-rooted. The deep-rooted plants actually bring moisture up, and the shallow-rooted plants feed off that in droughts. So we're able to get by on much less moisture than most people." Monocultures, on the other hand, have no root diversity, so the plants only access moisture from one level of soil, usually near the surface where the soil is driest. When fields contain shallow-rooted monocultures year after year, they exhaust the moisture resources at that level.

Because his pastureland is also diverse, Gabe can get by on less moisture there as well. Ranchers up and down the Great Plains live in fear of not having enough grass and hay for their livestock—and if they don't, they have to sell part or all of their herd. Gabe practically laughs in the face of drought. That's somewhat overconfident given the promise of future climate change–induced droughts, but he knows that his pastures can withstand extremes. They have before. "If you do a good job of rangeland management, one year of drought is never a problem. You don't have to destock or nothing because your soils have some resiliency built in them," he says. "I honestly believe here that we would have to get into the third, fourth year of drought before we start cutting back on any numbers at all. It's just not going to matter with the resiliency."

22

The Livestock

For twenty-six years, Gabe was in the specialty bull business, raising and selling breeding bulls, called balancers, or bulls that combine positive traits from several breeds into one animal. Being in the bull business meant timing the birth of his calves for February and March so they matured by sale season. February and March are cold, snowy, generally miserable months in North Dakota—awful weather for giving birth outside. Blizzards can dump several feet of snow on young calves, and temperatures can plummet well below zero. Calves born during a blizzard slide out of a warm placenta and land in a freezing snow bank—and their bodies go into shock. They shiver as the birth fluid creates an ice shell. Sometimes the tips of their ears and tails freeze and later fall off; sometimes they die.

My father's cows calve in March and April—not quite as early as Gabe's used to, but still challenging for rancher and calf alike. During blizzards, he races into the storm in his green and white '70s era Ford, seeking out chilled calves. When he finds one, he lays it on the cab floor and barrels toward the garage, where he places the listless calf on empty feed sacks in front of a heater. As a small child I would sit next to the calves and rub their cold ears and stroke their velvet faces. One knows a calf is thoroughly chilled when the air exhaled through his nose feels cold—so I dried the calves with towels and snuggled with them until, when I placed my hand over their noses, their breath felt warm.

It sounds like a nice scene: a child cuddling with a baby calf and a father heroically saving his herd from the storm. When the calves survive, it is. But blizzards and the sicknesses calves develop later take a number of lives each spring. Early-season calving is also hard on ranchers. I barely talk to my father during calving season because he spends those months in a sleep-deprived daze. On top of watching over the 450 or so range cows and calves, every two or three hours, day and night, he walks through the herd of 50 to 75 heifers (those giving birth for the first time as two-year-olds) in the corral. He can never leave the ranch for more than a few hours during calving season. When a heifer is ready to calve, he puts her in the barn. About once a week, he stays up most of the night helping one give birth. Ranchers put themselves and their cattle through this because they, too, want their calves to be as big as possible by sale time. A calf born in February has more time to grow than one born in May, and a heavier calf brings more money at auction from the CAFO buyers.

Gabe tells me that as his thinking grew more regenerative, he saw how the conventional system forces ranchers to do things that don't make sense, like calving in the winter, fattening their calves on grain, and producing animals for specialized breeders instead of consumers. He switched to raising grass-fed animals for meat and calving in late spring, when wild animals like deer, antelope, and buffalo give birth. Instead of barreling into blizzards, Gabe watches his cows calve comfortably on fresh prairie grass. Instead of corralling them and checking them every couple hours, he leaves them alone, which cows prefer anyway.[1] "We've gotten to the point where we don't keep cattle in the lots—very rarely, a little bit right after weaning," he says. "Now we start May 15 or 20 and we're done by the end of June. We calve out on pasture. Two miles from the nearest corral. If they have a problem, nature takes care of it. It's unbelievable. Our death loss is miniscule. We've been doing that five years, and we've lost one cow in five years. We've lost some calves, but you lose calves calving in the corral, too, especially in February or March."

I press him on this. I've watched my dad pull dozens of calves from cows that are having labor trouble—the calf is too big, the heifer is too small, or the calf is backward. Warmer weather doesn't solve an issue like that. He nods, having heard my concerns before. "People say, 'But what if you have problems?' You don't," Gabe says. "Nature takes care of it. We lose way fewer cows than we used to. It's just not a problem. I really think a lot of ranchers, we make our own problems. We think the cow needs help. If we just let her go, she'll work it out. Nature has a way of making things work out."

Gabe found that nature has a way of working out sickness, too. He stopped using vaccinations in 2009 (except for the Bangs vaccination). He's in his eighth year of not using dewormers, insecticides, fly tags, or any form of insect killer. He's noticed the same effect Phil did after quitting insecticides: a rise in the dung beetle population, which means manure is making its way into the soil much more effectively. When Gabe first quit using insecticides, he couldn't find a single dung beetle in the pastures. Now he's spotted fifteen different species, plus other predator insects that eat flies and fly larvae, such as dragonflies, spider mites, and Hister beetles. "It's just nature taking over and working. It's amazing," he says. "Do we have flies on our cattle? Sure. But it hasn't really been a big issue. You get a few cases of pinkeye, but those are the weak cattle. It's just like a weak plant. They'll die and the healthy will survive." Only a few cattle die per year of something that an insecticide or vaccination *might* have prevented. By keeping the offspring of the survivors, Gabe's herd grows stronger every year, which in turn means higher-quality meat for the consumer with no antibiotic or pesticide residues.

Gabe's hands-off approach extends into the winter as well: his cattle graze on open pasture and unharvested cover crops, a strategy most ranchers would scoff at. Ranchers in cold climates feed their cows hay during the winter. Every day, they start the tractor and shuttle bales to the pasture or corral. When the snow is deep, this takes hours. Ranchers feed hay because they don't have enough pasture forage to carry the cows through the winter; they grazed up their ranges during the summer and fall and didn't give them adequate time to recover. With his mob-grazing system, though, Gabe's land has plenty of forage to last the winter. The cows graze the native grass pastures well into December, then work on the cover crops until late January or early February. Gabe says ranchers are wrong to think cows can't dig through the snow to eat. Conventional cows don't do it voluntarily because eating hay is easier. Ranchers just need to remind the cows that they can dig. "We graze through two feet of snow all the time, it's no problem at all," he says. "They learn. The cattle I had ten years ago, could they do it? No. How we started this is, we just didn't feed and they learned. If they couldn't handle it, obviously we didn't let them die; we sold them and went on to the next one. The fittest survive."

But the cows aren't completely on their own, Gabe points out. His second winter strategy is bale grazing. Gabe dots hay bales across certain pastures in October. If the winter gets tough, Gabe puts the cows in the paddocks with the pre-positioned bales. Otherwise, the cows bale graze only from early February to late March or early April, depending on when the spring grasses return. The cows have free-choice hay, but they can still graze. Best of all, the system is labor-free during the cold months, a huge savings in time, fuel, and stress on the rancher. For example, Gabe's son, Paul, wintered 250 cows at his place last year, and he started the tractor just once to bring them bales. The only labor involved is positioning the bales in the fall and moving the cattle once a day or so (they move less often in winter than summer).

Driving along in the Ranger, Gabe shows me a paddock that he bale grazed last winter. I see a few circular patches of weeds where the bales once sat, but otherwise the paddock is as lush as the others. He points out one area that is full of radishes, great soil aerators. Weeds, radishes, and other plants that pop up where the bales were enjoy the high carbon content of the soil created by the hay breaking down. "It's just nutrients cycling through," Gabe says about the weedy patches. "Everybody says, 'You waste so much hay.' No, you don't; that's nutrients cycling. These annual weeds, they'll be gone in a year or two."

Gabe has just mentioned two of the most common objections to bale grazing: waste and weeds. Ranchers often believe that, given free-choice hay, cows will waste more than they eat by scattering the hay and trampling it until it's soiled. Hay is expensive to buy and time-consuming to make, so ranchers want to conserve as much as possible. The truth, Gabe says, is that the cows waste barely any free-choice hay, and not any more than they wasted when he used bale feeders. "Wasted" hay is not really wasted anyway, he says. When the snow melts, the trampled hay breaks down into the soil, adding nutrients for the next season. The weeds that come up where the bales used to be are simply part of that process, Gabe says. As he shows me, the weed patches give way to grass within a year or two. What's so wrong with a few weeds, Gabe asks, if the cattle will eat them? One doesn't want a pasture full of them, but a patch here and there won't hurt anything. "My philosophy is, as long as a cow or sheep will eat it, it's not a weed, it's forage," he says. "It's only a weed if they won't eat it. And there are very few things our cattle won't eat."

What about bad winters, I ask, when snowstorms lock people inside their homes for days, water tanks freeze up, and cattle suffocate under the snow. I think of October 2013, when a record blizzard struck western South Dakota. The Atlas Blizzard started with freezing rain that soaked and chilled the livestock; they did not have their winter coats yet and were still out in summer pastures because it was only the first weekend of October. The rain switched to snow, and it didn't stop falling for three days. Winds reached seventy miles per hour. Most weather stations had predicted light snow, so no one was prepared. Ranchers ended up snowed inside their houses with no electricity. Up to four feet of snow covered the plains and the Black Hills, burying cattle, sheep, and horses. Entire herds drifted into creeks and dams and drowned. Others huddled together, churning the ground beneath them into mud and drowning in the muck. About fifty thousand livestock died; ranchers in the hardest hit areas lost between 20 percent and 50 percent of their herds. Dead livestock covered the prairie for months—in road ditches, in creeks, under bridges, on Interstate 90. To make matters worse, the federal government was shut down and did nothing in response for several weeks. Somehow my parents only lost a dozen or so cattle; my dad credits the tree-lined ravines and gullies on our land that protected the cows and kept them from drifting.

I look around: there's hardly any natural protection for cattle here on Brown's Ranch, no draws or hillsides to huddle into. The land is flat and open. The only trees are in a handful of shelterbelts that he and Shelly planted years ago. Left out here in a blizzard, a cow would probably die. The mob-grazing system provides a solution, though. The portable fencing that Gabe uses to divide large pastures into small paddocks allows him to give the cows access to the closest protected area—a shelterbelt or a corral for example—if they want to get there. He builds in a corridor to safety. Cows can sense when storms are coming, and they will naturally find protection. I've seen my dad's cows do this: they bunch together against a cutbank or lie down in a deep ravine. The portable system also allows him to graze the pastures farthest from home in the summer and the land closest to home in the winter. Plus, a blizzard that isn't forecasted is rare, so if a severe storm is on the way, Gabe simply brings the cows home. The benefits of winter grazing outweigh the "what ifs," he says. "Cattle would rather graze than be standing in the corral," he says. "Living proof is in the spring and all the fences are leaning. They would. So why not allow livestock to do what they're evolved to do, which is graze?"

23

The Alternative to Hay

While Gabe enjoys growing cash crops like wheat, and of course his cover crops, he sees native grassland as the most important element of his operation. Every year more American grassland becomes cropland, and this is deeply unsettling for Gabe, a man who understands the world in ecosystems. Wildlife, insects, and soil microorganisms lose their habitat when grassland disappears. His view is straightforward: "I think we've ruined a lot of our northern plains with this grain farming. The good Lord didn't intend for all land to be farmed."

When we converted the prairie to farm ground, we destroyed its true wealth, which used to be underground in its deep, rich, tangled root systems. They were amazingly dense: one square yard of prairie turf just four inches deep can contain twenty-five miles of roots.[1] These roots protected the soil from erosion, flooding, and drought. Settlers ripped apart the whole system with a single plow pass. Today these root systems are all but gone, and the soil is rapidly disappearing as a result. As Dan Barber sadly points out, in *The Third Plate*, "The more you learn about the destruction of the prairie, the more difficult it becomes to see a modern wheat field as a thing of beauty, in the same way it is hard to see beauty in a clear-cut forest."[2]

Gabe seeks not to conserve the native prairie he has, but to create more of it—regenerating a resource, not sustaining a degraded one. North Dakota is a great place to start, since about 80 percent of the state's native prairie is gone, with the remaining portions mostly in the arid western part of the state.[3] The national picture is more dismal: just 3 percent of America's original 167 million acres of tallgrass prairie remain intact.[4] Gabe once had 2,000 acres of the farm in crop production, but in the last four years he has seeded more than 1,100 acres back to native grasses. "People think I'm nuts because we can make a lot of money grain farming, but that's not what I want to do," he says. "I'm kind of a livestock guy. I enjoy grass and livestock. We're going to make money. Will we make as much money? Maybe not, but that's not so important to me. We're evolving."

It's possible to coax native grass back with proper management, since many grasses remain deep in the soil, dormant until conditions are right for growing again. On cropland that has been farmed for generations, though, restoring native species usually requires seed because tillage has killed those dormant grasses. Where do you get native grass seed? I ask. Gabe tells me that some people were smart enough to preserve their native grass and go into the business of grass production, meaning they harvest the seed and sell it to people like Gabe who are doing restoration projects. Gabe bought his seed from a Minnesota farm with grasses adapted to the northern plains. He wants grasses that evolved to fit the environment, not imported grasses like crested wheatgrass. "I'm kind of a purist," he says, somewhat apologetically. "I want to see diverse grasses." Gabe wants diversity in his pasture for the same reasons he wants diversity on his cropland: diverse species build soil.

Gabe then mob grazes the native prairie. He uses portable electric fencing, called polywire, that he moves daily to create small paddocks that the cattle graze intensely. Usually Gabe moves the cow-calf pairs once a day and the grass-finished cattle multiple times per day. Here's how it works: The ranch has a number of large pastures with permanent electric fence boundaries, which Gabe divides as needed using a single polywire line attached to lightweight posts. Every morning, he creates paddocks that fit the size of the herd, the condition of the grass, and the lay of the land. Stringing the polywire is easy: he hooks one end to a permanent post, hops on the Ranger and drives forward, unspools the wire as he goes, stops every sixty to seventy-five feet, pushes a post into the ground with his foot, hooks the polywire to the post, and repeats. On nice mornings, he walks. When he's done erecting a section of polywire, he electrifies it so the cows won't break through.

It gets even easier—in his system, the cattle move themselves from paddock to paddock. Each polywire fence has a solar powered Batt-Latch gate release timer, which acts as a gate. Gabe sets each Batt-Latch to spring open at a certain time. When the time arrives, the Batt-Latch beeps and releases the fence, allowing the cattle to move to the neighboring paddock, where they graze until the next Batt-Latch goes off. The cattle learn what the beep means and readily move themselves to fresh grass. When the grass-finished herd is big and the grass is growing fast, they move six or seven times a day. "The cattle get so used to it that there's no work to it. No work to it at all," Gabe says. "Like for me in the morning, I walk down here and from the time I come down here, go out, roll out the polywire, check the water, mineral, all that: twenty minutes. It's simple."

When ranchers hear that Gabe moves cattle several times a day, they often shake their heads in confusion. They think it's too much work. I understand why: under the conventional grazing system, moving cattle is a big job, one that can take all day. If ranchers use horses (which my mom, sisters, and I did at our ranch), they have to catch them, saddle them, and ride or trailer them to the pasture. For ranchers using ATVs (my dad and brother), they fuel them and ride or haul them out. Then they have to find the herd, and because they use large pastures, the cows tend to fan out. They hide among trees or in draws, and it can take hours to round them up into one bunch. Finding them is only the beginning, though. Most ranchers move cattle a handful of times per year, so the cows don't always behave because they aren't accustomed to being moved. Sometimes they stampede, refuse to walk in the right direction, or break through fences, mixing with other people's cattle. Sometimes, for no apparent reason, they scatter in opposite directions like pool balls after the break. This is the work ranchers think Gabe is doing several times a day—and if that was what moving cattle on Brown's Ranch was actually like, it *would* be a ton of work.

Even when Gabe explains the system—the cattle move themselves; the polywire takes less than thirty minutes to set up—ranchers still object that his system involves too much hassle. Conventional ranchers want to turn the cows out to pasture and not mess with them for a couple of months except for checking water tanks and mending a fence here and there. Polywire and daily cattle moves seem like what my father would call "monkey business"—a waste of time and effort on an unnecessary project. Gabe doesn't see it that way. Under his regenerative system, he doesn't have to spend his whole summer haying; he puts up very little, mostly for the grass-finished cattle because he can't find nutritious enough hay to buy. He also doesn't spend days fixing barbed-wire fences or spraying crops, two jobs that suck up a lot of my father's time in the summer. "What the hell else do I got to do?" he asks. "Like this moving cattle. I love to do that. What's so hard about going and rolling up some polywire? 'Well, you could be busy doing other stuff.' Like what? I don't have spraying or anything."

"Better than sitting in the baler all day," he says later. I think about my dad and brother, who wake up at 4:00 a.m. and go to the hay field, where they use a tractor and baling machine to make hay bales out of the grass and sweet clover they cut into rows (called windrows) with the swather. These are the bales they'll feed the cows the following winter. Starting at age twelve, I went with my dad on these mornings and clambered onto the toylike red Farmall tractor. Its ancient engine stammered to life in the chill before sunrise as I looked for flat tires or broken teeth on the rake, a contraption with spinning wheels that whisk two windrows of hay into one fluffy, easier-to-bale windrow. When the tractor was warm, I engaged power to the rake, pushed the clutch to the floor (I was barely strong enough), jammed the stick shift into first gear, and released the clutch. I'd be off, the tractor bouncing over the rough field and the rake whirring behind me. Dad and I worked together baling and raking dozens of hay fields each summer, putting up thousands of bales. The summer after eighth grade, I graduated to the swather and spent ten to twelve hours a day cutting hay from June to August.

I miss being out in the hayfield and the peaceful solitude of a day spent gazing over the open land, occasionally spotting a newborn fawn or a disgruntled badger—and I *really* miss working with my dad. In reality, though, making hay involves hours of driving. Our hayfields are so big it takes days to cut them, then days to rake and bale them. It's a lot of sitting, like being a long-haul truck driver. It also uses a lot of diesel fuel. Think of a lifetime of making hay every summer like my father has done: that's thousands of hours on a machine doing work that isn't terribly rewarding, at least to my mind. When I think about ranch work that is rewarding, I think of the cows. If I were a rancher, I'd want to interact with them, not a machine. I realize now that my dad and ranchers like him wouldn't have to spend so much of their summer behind the wheel, a place ranchers seem ill-suited for anyway, if they changed their management style. The problem is clear: ranchers over-manage the hay and under-manage the grass.

Later in the afternoon, Gabe and I drive past an exceptionally beautiful pasture. Even from the road I can make out a few species: the brown flag-like tops of the bromegrass, the dark purple blooms of alfalfa, and the occasional bluish tint of big bluestem. I have to remind myself that it's not May but August, a time when pastures tend to look fatigued. I'm amazed by the pasture's diversity, but as a rancher's daughter I'm more impressed with the thick growth this late in the season. "I guarantee you most ranchers would want to hay that," Gabe says as we cruise by. He's right. The grass is just standing there with no animals on it; the cattle will either come later or have already been there. In the meantime, Gabe lets the grass grow as regenerative management dictates. Most ranchers would see the empty pasture and its gorgeous grass as wasted resources. They would be chomping at the bit to hay it, and I don't blame them—

that pasture would yield many tons of high-quality hay. During summers with abundant rainfall and cool temperatures, perfect weather for our ranch's cool-season dominant pastures, my dad will always hay portions of them, harvesting minerals without replenishing them. He's far from the only one who does this.

I call this urge to hay everything in sight the hay mentality. Like summer fallow, the hay mentality is an example of good intentions gone awry. Ranchers think that by haying their pastures they can store up extra feed in case of a bad winter or a drought, a move that protects their bottom line.[5] Plus, pasture-grown hay is a bonus: it appears without fertilizer and agrochemicals, making it input-free. Pasture hay is cheap from a dollars-and-cents point of view—all it costs is the fuel, machine use, and time required to harvest it—but, as Gabe explains, it is expensive from a soil point of view. "If we cut this now, you leave that ground bare, the temperature rises, and you are going to kill your soil biology," Gabe says. By now I understand the ramifications of dead soil biology: weak plants, less water infiltration, lower carbon levels, more erosion. For someone like Gabe who's trying to regenerate soil, killing soil biology by haying doesn't make sense.

Haying a pasture is even worse for the land than conventional grazing, and not only because of the lost soil biology. Ranchers can't graze that hayed chunk of ground because there's nothing for the livestock to eat, meaning animals won't fertilize it with manure and urine, aerate it with their hooves, or trample vegetation that will break down into carbon. Without organic matter to feed them, the grasses deplete their root masses, which in turn leads to the root die-offs that trouble Phil so much, which then leads to thinner, less diverse pastures. It's a classic example of robbing Peter to pay Paul. The rancher's bottom line benefits, but the land suffers—a quick savings with a long-term soil debt. In the end, neither the rancher nor the land is better off. "You've taken all that carbon that should be breaking down and feeding that biology, and you've exported it off," Gabe says with a sad shake of his head.

It's not that Gabe rejects making hay altogether. "We still put up some hay, because, hey, we're in North Dakota," he jokes. In all seriousness, there's no way around feeding some hay in cold climates, though he believes that if his land were contiguous and he could reach all his pastures during the winter, then he could forgo hay completely.[6] For now, he mostly hays land that can't support cattle—like road ditches or areas not conducive to the daily rotations because they are isolated—but he tries to cut as little as possible, and that's what he feeds the grass-finished animals during the bale-grazing season. He purchases what little hay is required for the cows. "With grass-finished animals you have to have really, really good quality," he says. "When they get to weighing a thousand pounds and you start feeding them poor quality, they are going to start losing condition. We can't have that. And I can't find high enough quality to buy. But all of our cow hay we buy."

As his pastures regenerate, Gabe has reduced his hay usage. "We've got it down now; the best we've done is only feeding hay 73 days a winter," he says.[7] That's no small feat. Conventional ranchers in North Dakota might start feeding hay steadily in October or November, depending on when winter sets in, and quit when the grass returns in April or May—at least 180 days a year, usually more. I think about the billions of bales ranchers like my dad have toiled to make, an unnecessary and even wasteful act when regenerative grassland management could mostly replace those bales. How many tons of carbon have they sacrificed? I wonder. How much of their land's life-giving nutrients did they deplete without restoring them? How long can we sustain such a system, and why would we want to?

24

The Restoration of the Native Prairie

Gabe and I climb into a pickup and head up the gravel road to see the cow-calf pairs. On the way he points out a field that he seeded back to native grass. "Everybody thinks I'm crazy," he says again. He keeps the cow-calf pairs separate from the grass-finished yearlings, with each group rotating through its own set of pastures. Right now, the grass-finished ones are near Gabe's house and the cow-calf herd is near Paul's. Four miles separate the two chunks of land, which makes it impossible to run the whole herd as one group. In those four miles are houses on forty-acre lots—too many yards, dogs, children, and hobby livestock to pass by with a herd of cattle. "How do you trail cattle by all this and the traffic?" he asks, waving at the houses we're passing. "It just isn't worth it. When the neighbor calls that your cattle are drinking out of his swimming pool, then you know you got problems. That's not a good thing." His dream is to own contiguous land so he can run the cattle together.

While it's been dry the last few weeks, overall the summer has been cool and wet—great for cool-season grasses like crested wheatgrass, an import from Asia that Russian agriculturalists brought back to their own prairies. Many years later, N. E. Hansen of the South Dakota Agricultural Experiment Station discovered the hardy grass growing along Russia's Volga River.[1] He sent samples to the United States in 1898. It wasn't until 1908, however, that seeds were planted on American prairies at South Dakota's Belle Fourche research station. More plantings followed in 1915 at a Mandan, North Dakota, research station not far from Brown's Ranch. By the 1930s ranchers were spreading crested wheatgrass across their ranges. This is one of the grasses Gabe is trying to replace with native grass. The going has been difficult because of the very traits researchers of the past loved about crested wheatgrass: its hardiness and ability to spread quickly. "A lot of people don't realize, with crested wheatgrass, its roots send off a compound that inhibits others species from germinating and growing," he says. "When you see crested wheat taking over, that's why. That's a defense mechanism that crested wheat evolved so it would be dominant. Why would we want to introduce something like that? Man does these things thinking it's for the best when really you're throwing nature all out of whack by doing it."

We drive along a fence line and Gabe points out the difference between his pasture and a neighbor's. The neighbor's land isn't completely bare, but the grass is short and thin compared to the tall, thick grass on Gabe's side. "He'll bring these up here at Easter and dump them in and leave them there until Thanksgiving," Gabe says, pointing at the cattle we see grazing on what looks like slim pickings. "They're just on the same pasture all the time. You wonder what those cattle think looking across the fence." I take a picture of the fence line comparison. I don't say this, but the grass on the neighbor's side looks about like my father's pastures do by August. "This is really good for him this year," Gabe says. "Normally you could play pool out on there."

Gabe turns on a dirt trail with grass between the tire tracks. He tells me we are driving past a field he bought in 1997 that the previous owner had enrolled in the Conservation Reserve Program. The whole field was nothing but bromegrass, he says. Smooth bromegrass, a cool-season variety that hails from Hungary, arrived in the United States in 1884.[2] Livestock owners planted it because of its drought tolerance and forage quality, which is higher than most cool-season grasses. Bromegrass spreads aggressively not only through seeds, but also through rhizomes, or underground stems that send out roots or new plant shoots (also called creeping rootstalks).

When the land was in the CRP, no cattle and few wildlife grazed it, which allowed the brome to take over. Gabe has been working for more than fifteen years to diversify that field. "Now through high stock density grazing, the alfalfa's coming back, we got sweet clover, we got some native species coming," he says. "It's still a lot of brome, but it's pretty good."

He points out some leadplant (*Amphoracanescens*) coming in, a nitrogen-fixing flowering shrub that grazers love. I recognize the leadplant, although I didn't know its name before—I see it all the time in my dad's pastures. As a kid I used to yank off its stalks of purple flowers and bring them to my mom. Maybe our pastures are a bit more diverse than I thought. Gabe mentions that just like in the cover crops, he wants flowering species in the pastures. "It's slowly getting better," he says, about the number of species he finds when he scouts the pastures. "We're getting a little more diversity, but it's not near what we want it to be. That's one thing I had to learn: you gotta have some patience. In nature one hundred years is a blink of an eye. We're only on the earth a short time and we want to see the difference immediately. That's not what nature's used to."

It's true: one hundred years is short given the long history of the Earth. Yet human beings have altered the landscape more in the last hundred years than natural forces like wind, rain, heat, ice, floods, or earthquakes have in the last thousand. Every animal, plant, and insect has had to adapt to our actions or die, and it's often painful to watch. Think of the lone Alaskan polar bear floating on a tiny ice chunk in the middle of an ocean heated by climate change. The natural world is in chaos, with some populations disappearing or decreasing under the changing conditions and others increasing out of control. Some people point to animals like the American crow and the mourning dove, two species predicted to increase in number as climate change progresses, and argue that climate change isn't inherently wrong because it helps some animals.[3] The crow and mourning dove are adapting and thriving, so they are obviously the fittest and therefore more deserving of survival, or so the thinking goes. Same with grasses like smooth brome.

But it's not really fair to make the natural world play by the rules of survival of the fittest when human beings have altered those rules so drastically. Life evolved slowly over millions of years, not quickly over one hundred or so years. Few changes occur hastily in nature; when climates evolve, or environments shift to different plant or animal species, or water temperatures fluctuate, the modifications unfold over thousands of years. A plant or animal might be perfectly capable of adapting over a long period, but could die if not given adequate time. So is it truthful to say a species doesn't pass the survival-of-the-fittest test and therefore doesn't deserve to be protected when we've sped up the evolution time?

Gabe's native grasses might be returning slowly according to the limited human conception of time, but the animals are already sensing a change and coming back. As the grassland grows more diverse, so does the wildlife population. "The amount of wildlife, it's just phenomenal," Gabe says about the ranges. "I like to use the grouse population as an indicator because it's a native species. We've got hundreds of grouse up here. Lots of pheasant, partridge, hawks, coyotes, foxes, song birds." Even the farmland supports wildlife. Conventional farming tends to make the land inhospitable to wildlife, with its lack of plant diversity, bare fields during winter, lack of fencerows for protection, and agrochemicals. In an article about the rapidly decreasing population of prairie birds, T. Edward Nickens writes, "Corn, scientists point out, is one of the worst crops for grassland birds. For starters, most species won't nest in corn, and corn production requires high inputs of chemical herbicides and fertilizers—bad news for natural ecosystems in general."[4] A bird population analysis of South Dakota, North Dakota, Minnesota, and Iowa conducted by the National Wildlife Federation found that grassland bird populations have decreased by nearly 30 percent in areas with high corn increases.[5] Birds aren't the only animals affected. Badgers, antelope, deer, raccoons, porcupines, rabbits, ground squirrels, coyotes, and mice are just a few of the species left homeless by corn production. With corn taking up an ever-increasing amount of real estate, there are fewer places left for animals to move to.

Gabe's fields provide a home for deer, antelope, numerous species of birds, mice, gophers, and many other animals. The difference between his land and conventionally farmed land is especially noticeable during the hostile Great Plains winters, when food and protection are scarce. "I can remember, when I first moved to that farm in 1983, you'd never see a deer on that home section. We would rarely see a pheasant or a grouse," Gabe says.[6] "Four winters ago, when we had 120 inches of snow, the Game and Fish Department came and flew over my place, and just in those cover-crop fields on that quarter with the tree rows—I had cover crops planted so my cattle could graze in the winter—they counted 876 deer on that quarter. They said they had deer that were tagged that came over fifty miles because they were starving. They needed something to eat. Well, where was there something to eat? The cover crops."

Many farmers would be annoyed to see deer munching on crops intended for cattle. But Gabe sees the deer as crucial links in the prairie ecosystem, of which his cover crops are a part. Like the cattle, tightly bunched deer fertilize and aerate the soil. Because his fields are part of the ecosystem, he can't deny access to wildlife, nor would he want to. Gabe does not see the land as his alone to use, but as something shared with the wildlife that live on it, a sentiment Dan Barber echoes when he writes, "Every farm is intimately linked to the larger ecosystem. . . . When you farm 'extensively,' you're taking in the world."[7] That is what Gabe does: he farms extensively, using the surrounding ecosystem to enrich his land and vice versa. The idea that farms are somehow separate from the environment and therefore off-limits to animals—and the related idea that farmers needn't yield anything to nature, even though farming the land robs resources from the environment—is a core tenet of conventional farming. Both are ideas that Gabe rejects.

We crest a hill and see the cows in a swale below. We are gazing over two sections of land (640 acres equals one section, so 1,280 acres) divided into forty-three permanent pastures fenced with barbed fire. Gabe and Paul divide those pastures into even smaller units with the polywire. I spot a water tank in the distance, and Gabe says he pipes fresh water into the units instead of relying on natural sources. Fifty or so water tanks dot the ranch, and as the cattle move he turns on the appropriate tank. "There are no creeks or rivers or anything through any of our land," he says. "It's either stock dams or fresh water. We still have stock dams in several, but we pipe fresh water into all of them. It works much, much better."

Earlier, I asked Gabe how he knows when to move the cattle. It depends on many factors: how much forage is growing in a pasture, how old the calves are (younger ones eat less, older ones eat more), what the weather is doing (cattle eat more during cold weather), and what kinds of grasses are available. He also relies on Brix readings of the forage, mostly for determining when to move the grass-finished cattle, which require the most nutrients. The Brix system, named for German chemist Adolf Ferdinand Wenceslaus Brix (1798–1890), measures the sugars and minerals dissolved in a liquid by looking at how the liquid refracts light.[8] This is done using a device called a refractometer. The angle of light refraction reveals the density and chemical composition of the liquid: the sugar and mineral content, or the nutritional value.

In the case of forage, the liquid is plant sap squeezed from leaves or stems onto the refractometer. Digital refractometers make it easy: put a few drops into the well hole, press a switch, and a Brix number pops up. The angle of light refraction corresponds to a number on the Brix scale. The higher the number, the more nutritious the plant. Oranges grown in nutrient-poor soil might clock in at eight, for example, while those grown in rich soil might read at twenty or better. Winemakers use Brix readings to determine the nutrient density of their grapes, which in turn affects the taste of the wine. Nutrient-rich grapes equal delicious wine, and forage is no different. In fact, cattle prefer such forage and can tell the difference between nutritious and non-nutritious grass through their facial hairs.[9] These hairs relay nutrition information to the cow, who then decides whether the grass in front of her is worth her time and energy. If not, she moves on. Nutrient-dense forage also boosts the cow's health, which is especially important to Gabe because he uses no chemicals to control insects or diseases. Healthy cattle resist sickness and—here's the part my consumer side is interested in—they produce better-tasting meat. Just as nutrient-dense grapes make good wine, a nutrient-dense steer becomes succulent steak, ribs, and burgers that not only taste delicious, but also provide more nutrients. In addition to the bovine benefits, using the Brix system helps ranchers use grassland resources more responsibly. Nutrients are consumed at their highest levels and don't go to waste, which means ranchers can use less grass to feed more cattle.

That's exactly what Gabe is doing. He moves the cattle to new pastures when the Brix readings are high, forcing them to eat the grass at its nutritional peak. He has discovered that Brix readings are usually highest in the afternoon because plants have had more time to photosynthesize. He might let the cattle linger in a pasture during the morning, working up an appetite, then move them several times in the afternoon over high-nutrition grass. He harvests hay using the same philosophy: he never cuts hay in the morning, always in the afternoon and evening so that it contains more nutrition. The Brix readings don't stop there. Gabe tests his garden vegetables and any produce he buys at the grocery store or farmers' market.[10] "You want to create a stir, go around the farmers' market testing Brix on everything," he says with a chuckle. "They won't like it."

He rummages around the machine shop looking for the refractometer. After a few minutes he finds it sitting not far from the chicken evisceration table. He tears off a few blades of grass growing near the shop doorway, squeezes green juice from them with a vise grip, and drips the liquid into the handheld refractometer. Gabe's refractometer is an optical model, not digital. It looks like the tube-shaped eyepiece of a microscope that has been snapped off. He peers down the tube to look at the prism, where there is a thin layer of plant juice, to determine the angle of refraction and thus the density. A scale embedded inside the viewfinder reveals the Brix number.

The grass reads between three and a half and four—very low, Gabe says, shaking his head. Ideally, he tells me, plant health starts at around twelve. A crop at twelve or higher will resist most disease pressure, insects, and weeds.[11] I am confused. To me, the grass appeared nutritious: dark green, thick, and tall. I would have turned animals into a field of it thinking they'd fatten up. Gabe explains that the soil around the machine shop doorway is poor, meaning the grass has little access to nutrients. It might look healthy on the outside, but contains no nutrition on the inside. "The goal is to get your soils healthy enough that you have a higher Brix reading," he says. "It's not uncommon for us in our alfalfa and some of the grasses we put up for hay to have eighteen to twenty-four, really high."

I ask Gabe how he learned about the Brix readings. Late at night on the internet, he says. "One of the things about me is, I don't sleep a lot," he says. He tells me later that a good night of sleep for him is four hours; he averages three. He spends most of his sleepless nights reading and researching. He reads two or three books a week, more than I do, and my job is to write them. Talk about combating the dumb farmer stereotype. Reading is yet another thing Phil, Kevin, and Gabe have in common. I'm starting to think I should add "reading and other forms of self-education" to the list of components in regenerative agriculture. "My mind is always going. I'm not a scientist; I don't need to know all the intricacies and all the big words, all I want to know is what works and let them explain to me why it works," he says. "I'm always studying and trying to learn."

After looking at the cow-calf pairs, Gabe and I get back in the Ranger and cruise out to see the grass-finished cattle. On the way we pass the chickens pecking at the grass in front of the modified horse trailers Gabe calls eggmobiles.[12] He's renovated the horse trailers to include roosts and waters and a tiny door. A loose fence made of netting surrounds the trailers so the chickens have a "pasture" to graze. When the chickens exhaust their pasture, Gabe hooks the eggmobile to the Ranger, pulls it to fresh ground, and sets up a new grazing area. One eggmobile houses broilers, or chickens grown for meat. They follow the cattle and "clean up" after them by picking at the insects gathered in the manure. When needed, Gabe feeds them peas or oats grown on the farm. The broilers have fresh air, natural food, and plenty of room to walk around, quite an upgrade from the crowded chicken confinements most broilers live in. "That will be real meat that you're eating there," Gabe says, pointing at the pecking broilers.

Another eggmobile houses pullets—young hens less than a year old—that just started laying their first eggs. A third eggmobile is home to the laying hens; these hens are so used to the system (and so concerned about staying close to their eggs) that they don't need the fence. Every day or two Gabe hooks up the eggmobile and the hens follow along. "Talk about no work," he says. This winter the chickens will stay in a new hoop house that Gabe is planning to build in the fall. In years past, they weathered the winter in a barn with a run that led outside.

We stop to look inside an eggmobile, bending over to see through the tiny door. "Girls, we've got a visitor!" Gabe yells. The hens peck at the floor, unconcerned. He sells their eggs at the farmers' market. "The longest it's taken us this year to sell out is about ten minutes," he says. "They'll stand in line over an hour. Four dollars a dozen and people just go nuts. It's unreal. There is a huge difference between a factory egg and an egg from here." I've tasted the buttery, bright yellow yolks of farm-fresh eggs and I understand why people go nuts. They taste like *eggs*. We jump back in the Ranger and keep driving. "Okay, girls, you gotta move," he yells at the hens, not without affection. "They get like pets when I come out to do chores. They jump up in here." I chuckle, imagining Gabe driving around with a hen in his lap.

As we drive, Gabe tells me about the morning he saw a coyote sitting right by the eggmobile door. Gabe was at the breakfast table and could see the eggmobile from the window. Its aluminum door has a photosensitive eye: when the sun goes down, the door closes (the hens know instinctively to roost before nightfall so they're already inside), and when the sun comes up, the door opens. The coyote had learned that the door's opening coincided with the sun and he was waiting on schedule, poised to nab a hen the second she stepped out. Gabe leapt from the table and burst out the door, but the coyote grabbed one before he could get there. He loses twenty to twenty-five chickens a year to coyotes and other predators. "People say, 'Well, why don't you kill all your coyotes?' If I start killing all my coyotes, then we're going to start having gophers to no end, rabbits to no end," Gabe says. "It's all part of the ecosystem. You can't do that." Just like he won't chase the deer from the cover crops, he won't shoot the coyote that eats a chicken now and then. Most ecosystems include predators to maintain balance in prey populations, and an unbalanced ecosystem will collapse. The coyote is simply doing what it evolved to do: consume prey. Gabe doesn't fault the coyote for that, even if it means losing some chickens, because he realizes he can't keep all of nature's bounty for himself.

We reach the grass-finished herd, which consists of steers and heifers borne by his cows. Unlike conventional ranchers who wean in the fall, Gabe leaves his calves with their mothers all winter and weans them at the end of March. "The goal is to have as little expense into them as possible," he says. "We don't have to have them in a lot where they're getting sick and we have to start a tractor to feed them every day. They're better off just out on the cow." I ask if the cows lose weight over the winter, being pregnant and still nursing last year's calf. "The cows are getting better," Gabe says. "They'll lose a little body condition during the winter, but right after we wean the calves end of March, first of April, as soon as the grass greens up they go out and they just blow up like ticks, put on weight."

Gabe keeps the weaned calves and runs them another year on grass (they are termed "yearlings" at that point). The heifers are exposed to bulls over the summer; those that breed rejoin the herd of cows and give birth the next spring. Gabe sells most of the yearling steers to companies that slaughter them and market the meat as grass-fed. When the remaining heifers and steers become two-year-olds, they enter the grass-finishing stage, after which Gabe will have them slaughtered locally. Instead of fattening cattle for slaughter using CAFO feedstuffs like corn or ethanol by-products, Gabe uses grass, proving, like Phil, that CAFOs are not necessary for meat production.

Most of the cattle we're looking at are two years old, all are heifers, and several will be ready to harvest in the next two months. Gabe tells me he prefers to grass-finish heifers because they fatten better and faster than steers. He moves this bunch three to four times a day (well, the Batt-Latch moves them). Gabe shuts the Ranger off, gets out, and walks slowly toward the herd. I follow, not looking at the cattle yet but at the grass under my cowboy boots. The pasture looks nothing like the one Gabe said ranchers would want to hay: the grass is trampled, the soil is churned black in some spots, and manure patties dot the landscape. That's how it's supposed to look, though, when grazers impact the land intensely. Instead of being scattered across a several-hundred-acre pasture, these heifers are clustered together on several dozen acres, tightly bunched like a herd of buffalo or antelope would be. Some have slick black coats, others have shiny red coats, and one oddball has a creamy white coat. One steps forward from the herd to inspect me more closely.

I try to imagine these heifers in a feedlot. They would be standing on cement in ankle- or knee-high manure and urine. They would be eating corn-heavy rations that make their rumens malfunction and livers shut down. Their waste would flow into lagoons that become toxic pools of sludge. That waste would spoil the air in nearby communities and the water in nearby streams and rivers. The heifers would sleep in that waste until their hair became matted with it. They would become less like cattle and more like corn-processing machines understood in terms of pounds gained per day compared to feed ingested.

At Brown's Ranch, the heifer is not a machine. She is part of an ecosystem. Her worth is quantified not just in the pounds of meat she yields, but also in the pounds of organic matter she returns to the soil. She is prized for her ability to feed not only people but also soil—which in turn feeds microorganisms that feed plants that feed other animals, including more cattle. She produces the same amount of beef as a CAFO-raised animal, sometimes more, without the costly inputs. As with cash crops, the Browns have discovered that it doesn't take inputs to produce meat. What it takes is healthy soil.

The Browns have eliminated not only the CAFO and its inputs, but also the industrial slaughterhouse. "The next step we did—it took a long time—but a bunch of us built our own slaughter facility at a small town called Bowden," Gabe says. Out of necessity, the Browns, their like-minded neighbors, and other investors joined forces to fund and build the small slaughterhouse, a move most ranching communities wouldn't think to make. "We are part owners," he continues. "Paul is actually treasurer of the board. So we slaughter our grass-finished beef, lamb, and pastured pork up there."

He describes the time-consuming inspections and paperwork involved with building the plant as "a huge headache," but after three years the facility is finally done. "It can only handle fifteen head a week," Gabe says somewhat regretfully. I say I'd rather eat beef that comes from a fifteen-animal-a-week, locally owned facility than a five-thousand-animal-a-day megaslaughterhouse. He says that's one reason he and his neighbors decided to build the plant: people are growing uncomfortable with the harmful social and health effects of megaslaughterhouses, like worker exploitation and meat contamination.[13] They want their meat not only raised responsibly, but also slaughtered responsibly. Gabe already feels the pent-up demand. "The plant just got up and running in January," Gabe tells me.[14] "We thought, boy, if we could sell one to two animals a month! Well, we're selling one to two a week, and we can't even begin to keep up. It's just unreal."

Gabe explains that North Dakota had just a handful of state-inspected slaughterhouses before the Bowden plant opened. The demand at those plants was prohibitively high, Gabe says. "The waiting list to get something harvested was over a year out. For us, trying to run a business a year out was . . ." He shakes his head. "That's why we decided we gotta invest and do it." The plant cost $1.3 million to build—and it was completed debt-free, Gabe is proud to add. Being a key investor and the plant's largest customer, Gabe has several days a week "locked up" for processing his own animals. Now he is able to bypass wholesalers and market meat at the local farmers' market under his own label, Nourished by Nature. "This is much more rewarding," he says about the transition. One reason is that he knows the meat in the packages is from his animals. "With a wholesaler it gets mixed," he says. "You don't know what's yours." I think of Ryan Roth and the vegetables that get packed in the Roth packinghouse under a number of labels from a number of different farms. The only way Ryan can be certain that he's eating his own produce is if he handpicks a head of lettuce or a sack of green beans from the field. It's the same with my father's crops. The truckloads of wheat he sells are mixed with wheat from thousands of other farms and ground into nameless, traceless flour.

The second reason selling meat is more rewarding than selling live cattle? The transition introduced Gabe to a person he rarely encountered before: the consumer.

25

The Farmers' Market

I didn't visit a farmers' market until my early twenties. It wasn't that I didn't want to, it was that I never lived close enough to one until I moved to college. I grew up about as far away from the reach of fast food and grocery chains as you can get in America, yet there was nowhere for my family to buy locally produced food, at least not in an organized, dependable way. If you knew someone who had chickens (a rarity even for such a rural place), a milk cow (even rarer), or a summer garden, then perhaps you could buy some food here and there. Maybe you could call rancher neighbors and ask if they had any beef for sale. But there are no regular markets or even a farm stand. Nothing in Bison's tiny grocery store is local. Nothing on the menu of the town's one restaurant is locally raised. The corn in the surrounding fields isn't edible until it becomes corn syrup and other by-products, or until CAFO-raised cattle eat it and turn it into fatty, dull-tasting meat. Practically none of the big farmers keep barnyard animals like hogs, chickens, or milk cows. Most don't plant gardens. Many of them raise wheat, but have never eaten freshly ground flour, let alone milled wheat themselves. They taste none of the fruits of their labor until a big company returns them in the form of processed food. Even in a place where cattle outnumber people thirty-five to one, people eat CAFO-produced meat.[1]

Such a situation is an inevitable consequence of industrial agriculture. Farmers are now dependent on the sprawling food system they helped create and lack the power to feed themselves. Eating local is even harder than eating organic in many rural communities like Bison. Even if farmers reject the industrial production system, they still have to eat within it. Or do they? Gabe would argue no, that farmers hold the key to changing the system if they would put their minds to it. They have to find or re-create markets that disappeared as a result of conventional agriculture. "I tell people, what was it like in Missouri, for instance, one hundred years ago?" Gabe says. "You can't tell me it was just corn and beans. We've lost these markets because of the current industrial production model, which makes no sense."

It's true: as farmers specialized, so did food markets. Midwestern grain elevators that mostly handle corn, soybeans, and wheat today once traded in a variety of small grains. Local slaughterhouses used to process a community's livestock—chickens, geese, hogs, sheep, and goats, not just feeder cattle all at market weight. The push for uniformity and transportability, though, shifted demand away from localized crops and toward standardized plants and animals. Regions built markets that can handle only a few crops. In Iowa, for example, the market has been transformed to handle mainly corn, soybeans, and hogs. Farmers there would have a hard time selling, say, a crop of flax. Elevators would balk at it, and few specialty chefs or food processors exist there to buy it. Iowa is not the only region with a specific market: think of Idaho's potato fields, New York's dairy farms, and Florida's winter tomatoes. Difference is not celebrated in today's standardized food markets, but rejected and, in many cases, tamped down.

Specialized markets have created what we might call the corn and soybeans ideology. Midwestern farmers often claim they grow strictly corn and beans because that's what the market demands; anything else won't sell, they argue (never mind the fact that because all the farmers threw their lot in with corn and beans, the market followed suit, not the other way around). All over the country, farmers fear they won't be able to sell alternative crops, so they don't try. Gabe has little patience for that argument. "I say, does your semi have tires? So what if you got to go an extra four hundred miles? You can find the markets." Gabe offers an example of farmers he met in western Nebraska who wanted to grow peas to help boost soil fertility. The problem was that they had nowhere to sell the peas after they harvested them. These farmers created their own market by convincing nearby cattle feedlots to buy the peas for feed. Perhaps this example isn't perfect—in an ideal agricultural system, we wouldn't have feedlots—but the point is clear: farmers can be creative in finding a market for alternative crops. Gabe rejects the idea that they are trapped by current markets and instead argues for the farmer's freedom to find or create new ones, independent of conventional agriculture's demands. "There are ways to market," Gabe insists. "You can get several people together. Maybe you have somebody closer to a metropolitan area that you can work out an arrangement with to sell the products, but there are ways."

Gabe has a ready-made market, Bismarck and the surrounding metropolitan area, practically in his backyard. He doesn't have to go an extra four hundred miles to sell his grass-fed beef. I can't help but see how easy it is for Gabe and Kevin to sell alternative products when urban customers are within easy reach. I think of Phil, though, who lacks a market for retailing grass-fed buffalo meat. He's where Gabe was a year ago, selling all live animals because he has neither a slaughterhouse nor a way to reach consumers. Is Gabe's argument wishful thinking for isolated producers like Phil, or my brother for that matter? I don't think so. Ideally, the markets will appear or at least get closer as regenerative agriculture takes root. This kind of agriculture will revitalize rural communities, which will give farmers and ranchers more opportunities for marketing. If we can rein in our top-heavy food system and replace it with a regionally and locally driven system, then farmers will likely enjoy more demand from places now closed to them, such as retail stores, restaurants, schools, farmers' markets, food processors, bakeries, anywhere people buy food.

Gabe sees these kind of farm-to-consumer transactions as empowering for farmers. When farmers and ranchers set their own price, something they cannot do under the conventional system, they can better ensure their economic survival. "We're capturing all that extra value," Gabe says about selling retail meat instead of live cattle. "Right now, a fat animal will bring about $1.60 a pound, say, if it was in the conventional market. A 1,200-pound animal, you're at whatever that comes to be, $1,800, $1,900, something like that. We are making much, much more," he says. He tells me exactly how much more, but, as midwestern modesty dictates, asks that the number not be put in print. The point is not how much profit Gabe makes, but the fact that vertically integrated companies like Cargill, Tyson, and JBS don't control him. He doesn't conform to their demands, but to his customers' demands. Here, the consumer has real power. The price Gabe charges better reflects the meat's true cost while simultaneously removing the social, environmental, and health costs of industrial meat. Gabe would rather see responsible ranchers empowered, not corporate beef producers. "Us as farmers and ranchers, we're used to giving somebody else all the profit," Gabe says. "Why not take it for yourself? Then we have control. What's wrong with taking the next step and putting more of those dollars in our pocket?"

We walk into the pull-type concession trailer that's parked in the farmyard. It's a white, rectangle-ish trailer with a flip-open counter, and the words "Nourished by Nature" and pictures of cattle, sheep, hogs, and chickens out on pasture are emblazoned on one side. It looks somewhat like a food truck. I stare into a deep freezer full of packaged beef, lamb, and poultry, the waft of cold air refreshing in the August heat. A refrigerator sits in another corner; that's where Gabe keeps the eggs. I open it and gaze inside, mostly for the cooling effect, though I claim to be admiring the eggs. Three days a week during the summer, Gabe and Paul work the farmers' market in Bismarck. In the winter, when the market is closed, they email order forms to customers, who meet them in town to pick up their goods.

Buzzwords like "grass-fed" and "local" might attract customers initially, but Gabe is more interested in convincing people to buy his products because of the regenerative agriculture system they represent. "What we're trying to do, then, is teach our customers that we're doing the full circle,"

Gabe says. "They come and the first thing they want to know is, where are you from? And we tell them. Then they want to know, GMOs or not? That's number two. Then it's antibiotics, hormones, and grass-finished. Those questions follow. What we're trying to do is sell the whole story. We're trying to produce healthy food on healthy soils, and hopefully that will equate to human health." Despite his preacherlike vibe, Gabe isn't on a mission to convert, but to educate. It's up to consumers to make the choice once they have the facts, he tells me. "I'm not going to get in an argument with anyone about GMOs, non-GMOs, all that kind of stuff," he says. "All I'm saying is, this is what we want to offer. If you want it, fine. If not, that's fine. Everybody should be allowed to put in their mouth whatever they want as long as it's legal. It's just our choice that we would rather not have GMOs. Hence the name, Nourished by Nature, our trademark. We want it to be as natural as possible. We don't want any of the synthetics."

I try to picture my father interacting with customers at a farmers' market and, well, I can't. Retailing isn't part of the modern rancher identity—it doesn't jive with "get big or get out" because it takes time and resources away from production and expansion. I ask Gabe if he ever thought he would be marketing meat directly to customers from a trailer like this. Never, he said. "It was never a real desire of mine, but as I've grown and studied more about human health, it has become that way," he says. The connection between non-nutritious food grown in dead soil and poor human health is all too clear to him, though he admits a part of America's health problem stems from bad food choices. "You can't blame it all on poor soil. A lot of it is eating habits—fast food, processed foods," he says. The problem is that, even when consumers chose vegetables and grains over fast food, their health still suffers because the "good" food doesn't contain the nutrients it should. The way to fix this problem is to grow food in healthy soils, plain and simple. "A lot of it, I think, is what we're doing to do our soils," he says. "If we can teach people, even a little bit, that healthy soils can produce healthy food, that's what it's about to us."

It's the soil, stupid. I imagine this must be a tough revelation for Gabe to bring about in customers. The mental leap between soil and food seems easy to make, except that powerful agribusiness voices keep insisting there is no connection. Even I didn't grasp the importance of soil until I started writing this book, and I went a bit deeper into the subject than most people go. How do we help urban consumers, who've lived off the farm for generations, value soil that they might never hold in their hands, never gaze over in the spring, never plant seeds into? How do we insist that they help protect microorganisms that they can't see? I'm not saying urban consumers can't grasp these concepts—far from it—but it's been extremely difficult even to get farmers, who live on the land and rely on it for their economic survival, to care about soil. Yet without a focus on soil, the regenerative agriculture movement can't succeed.

The customers in Bismarck, at least, seem to be making the connection between soil and food, as shown by the demand for Gabe's products. This winter, he hopes to buy a walk-in freezer; the deep freezer is already too small to accommodate his sales. He is also reaching out to chefs to develop a line of Nourished by Nature sausages without nitrates, nitrites, or GMOs. That's about as far as he wants to go with chefs, though. He has received requests from restaurants, but he has turned them away because they want choice cuts and nothing more. "They want all T-bones or porterhouses, and what do you do with the rest?" he says. The demand for choice items is a major problem in the farm-to-table movement: farmers end up growing popular, marketable crops or cuts of meat that are often ecologically taxing. As Dan Barber argues, "Farm-to-table chefs may claim to base their cooking on whatever the farmer's picked that day (and I should know, since I do it often), but whatever the farmer has picked that day is really about an expectation of what will be purchased that day . . . the farmer ends up servicing the table, not the other way around. It makes good agriculture difficult to sustain."[2] Barber is wise to point out this flaw in farm-to-table production, and his observation applies to farming in general. Unlike most cuisines, American food culture revolves around food fads and choice items. We don't pride ourselves in using the whole animal or cooking with edible but unorthodox crops.

It's not just chefs who demand choice cuts—most shoppers do, too. People prefer cooking with familiar items like steaks and ground beef. That's one reason we have CAFOs: so everyone can eat steak and avoid what most of us consider gross or unrefined items like brains, tongue, and liver. A quick scan of any grocery-store meat counter confirms this. In Bismarck, some customers are more adventurous: "Soup bones are one of our best sellers at the farmers' market," Gabe tells me later. "Liver also sells. There is a market for everything, including the fat." These consumers are anomalies for the most part, though. Gabe says that people have lived in the convenience-food era so long that many don't know how to handle basic cuts, let alone more adventurous products. "They have no idea how to make a roast. It just blows our mind," he says. "We sell ground beef in chubs, it's in a package. They look at it and say, 'No, I want hamburgers.' They don't put two and two together. In Bismarck, North Dakota, you'd think . . ." He shrugs and shakes his head. "But it's not that way."

Eating high on the hog, so to speak, is inherently unsustainable, even if we use regenerative agriculture to produce those choice items.[3] If we want diversity on the land, then we need diversity on our plates. Remember Ryan Roth's attempt to grow hybrid radishes that looked unfamiliar to consumers, and how he couldn't sell them? Consumers will have to reject standardization at the grocery store and be open to new things. Changing a nation's food culture sounds daunting, but think about how rapidly our food tastes changed with the appearance of processed foods and industrial agriculture. Think about how quickly we forgot what a vine-ripened tomato tastes like and came to see green tomatoes ripened with gas as normal. Think about how often people cook a whole chicken versus the breast only, or how we reach for a can or box instead of making something ourselves. We radically changed our conception of good food in less than a hundred years. If we've done it before, we can do it again.

This year, the Browns started an internship program, with the long-term intention of helping a new generation of farmers learn to practice regenerative agriculture. They welcome young people from any area: urban, rural, East Coast, West Coast, and anywhere in between. The first season hasn't gone exactly as Gabe expected. Of the hundred or so applicants, the Browns selected two young men and one young woman. The young woman didn't stay long; she was hired to help with marketing but later revealed that she "didn't like people," which made a public relations role difficult. Of the young men, one is the son of Gabe's farmer friend from Indiana. When the twenty-one-year-old arrived, Gabe discovered he didn't know how to use basic tools—tools so rudimentary I can use them, like vise grips—or how to change a tire (again, something even I can do). Gabe politely says that the learning curve has been high with this intern, but he's improving. The other young man, Troy, is a better fit, partly because he worked for the Browns during previous summers. But mostly, Troy is a hard and conscientious worker; he notices things that need to be done and completes them without being told.

Despite the initial setbacks, Gabe is optimistic about the intern program and plans to keep it going. He believes that rural areas aren't the only legitimate places for farmers and ranchers to grow up, that urban people can and should return to farm country to help revitalize food production. These young people need guidance and time to attain skills that farm kids grow up learning, he says. "There is a huge number of young people that want to come back to the farm, but they're not from the farm, you see," Gabe says. "I really think the next generation of true agricultural leaders is going to be from the city because they are way more open-minded and they know how city people think and what they want. I say 'city people' kindly."

In fact, Gabe would provide start-up capital and even land for the right young person. "One of our passions is to help young people. If the right interns come along and they really have a passion, too, I'll get them started. I'll buy the land and get them started. Paul and I have talked about it," Gabe says. "We would love to have somebody come and intern and raise bees. We would love to have one start with rabbits. We would love to grow the orchard and the vegetable business. But they gotta have the drive and want to do it. I'm not going to have them say they're going to and come and then all the sudden they don't like it. If the right young people come along, I'll help them get started."

I am amazed by Gabe's frank admission that he would help a young farmer in a significant financial way. I worry that he's a bit overconfident in the next generation and will grow disappointed with the interns over time, or that he will invest in the wrong person. I think of Eustace Conway as profiled in Elizabeth Gilbert's book, *The Last American Man*. Eustace lives off the land in the Appalachian Mountains, wearing clothes made from animals he's skinned and eating food he forages, kills, or grows. He rejects capitalism, consumerism, and "boxes," which are televisions, cell phones, suburban houses, and cubicles. He started an internship program for many of the same reasons Gabe did: to teach the next generation how to live from the land. Over and over the interns disappointed him. They wouldn't work hard, or couldn't handle the lifestyle, or weren't truly committed to it. Granted, what Eustace asked young people to do—give up the materialism of American life and find meaning in nature—involved a radical change. Gabe will likely have more success teaching young people how to farm regeneratively, but I wonder whether those teachings will translate into actual farms. The right person might seem right until money enters the picture.

Besides, shouldn't those resources be used on Brown's Ranch or maybe saved for future grandchildren? I ask. That's not how Gabe sees it. "Spiritually, my wife and I and son honestly believe that God put us through those four years so that we could get to this point. I'm just paying it back," Gabe says. "I saw the difference it made for us. I'm just trying to help people as much as I can. When you're as broke as we were for those years, you realize money doesn't matter. It's nice to do something once in a while, but money to me, it doesn't drive me. I could care less. I could sell this place for millions and not do a thing, but what good does that do? What do I do? It's not me. To me, it's more important if I can help other people and help the next generation, and inevitably I really hope we can change the production model and improve human health."

The next generation on Brown's Ranch is, of course, Gabe's son, Paul. Gabe says he knew when Paul was a young boy that he would grow up to be a farmer. Paul loved being outdoors and helping with farm chores, but he also understood the importance of regenerative agriculture from an early age. Plus, their farm is a fun environment that young people would want to return to, Gabe says. Not that there aren't hardships, such as animal deaths or equipment-related frustrations, but Brown's Ranch isn't like conventional farms and ranches, where people race to get big or get out. That's a key difference between conventional and regenerative farms: there's room for personal fulfillment under the regenerative model. "What's so hard about this life?" Gabe asks, referring to life on his farm. He answers his own question yet again. "The money is there. The joy is there. Why wouldn't you want young people to take over? I get really frustrated when I'm out speaking and I see [farm] couples who have children, but they discourage them from coming back [to the farm]. That bothers me. Why don't you want your children there? To me, that's a compliment if your children want to take over the business. That's how it should be."

Paul graduated in 2010 from North Dakota State University (NDSU) with a major in range management and minors in crop and weed science and animal science. Gabe tells me that Paul found college far less interesting than life on the ranch. The agriculture NDSU teaches is not regenerative, but industrial, meaning Paul saw most of his courses as a waste of time. The value of his degree lies not in its agricultural lessons, but in that it allows him to teach at the university level. For a while he taught at Bismarck State College. Gabe says Paul enjoys teaching because he wants to help young students see that there's a different type of agriculture available than the one taught in most universities.

Most semesters on the first day of class, Paul asks his students if they are from farms and ranches; the vast majority always is. Then he asks if any know the names of the consumers who buy their farm's products. Paul has yet to have a student raise his or her hand. Most of them come from conventional farms, so Paul has the difficult task of introducing them to regenerative agriculture and helping them understand why it's better than industrial—a lesson many students balk at because it inherently challenges the way they were raised and the work their parents do. I understand why some students reject Paul's philosophy; I rejected such ideas for a while, too, for the same reasons. I did not want to admit to my family's participation in a flawed system, so if I refused to believe the system was flawed then I could avoid the guilt. In many ways, the students' reactions mirror the reactions of beginning students in any course that challenges their preexisting notions. In the writing classes I teach, for example, students from conservative households often argue that racism does not exist, or that poverty occurs strictly because people don't work hard enough. When we look at evidence that proves otherwise, some get angry and retreat further into their preconceived notions—but most begin to open their minds and engage in the type of critical thinking required in today's world, and many of Paul's students do the same.

Like Gabe, though, Paul realizes he can't reach everyone. All he can do is expose students to new ideas and let them make their own decisions and create their own agricultural philosophies. He takes them on field trips to see how people are practicing various forms of regenerative agriculture, including to Brown's Ranch. Every year, Gabe tells me, one or two students truly understand what Paul is teaching and "run with it" by starting their own diversified and holistic operation. Like his father, Paul is making a difference for the next generation, equipping them with the skills to succeed in a world that, I hope, will increasingly demand regenerative agriculture.

26

The Message to Conventional Farmers

Gabe and I sit and visit in the machine shop, a gray steel-sided building filled with tools and machinery. It smells like grease, just like my dad's machine shop. He opens a beat-up refrigerator to grab us bottles of cold water and I spot tempting craft beer inside. A blind border collie, Pistol, rubs against my leg, asking me to pet him, which I do. I tell Gabe I'd like to talk about the controversial stuff, the political stuff, the challenges. What's it like to speak to auditoriums full of farmers who think you're a crackpot? Gabe leans back in his chair, ready for a long-haul conversation.

"Everybody laughs and makes fun of you, says you're going to go broke and all that. We just laugh." Gabe actually laughs at this point. "You can't let it bother you. One thing about me going out and speaking—and I don't know how many people I speak in front of every winter, thousands upon thousands—you always have your naysayers. You realize only a very, very minute percentage of those are going to grasp regenerative agriculture right away. Some will grasp it a little further on. Even so, 95 percent of them will never get it. So be it. I can only do what I can do."

Farmers and ranchers fail to understand the regenerative model not because it's too complicated, he says, but because they don't want to. They're wearing the conventional agriculture blinders, like some of Paul's students or the family in Colorado. "There's an old saying that goes, when the student is willing, the teacher will appear. You've gotta have that philosophy," Gabe says. "I tell people, I'm not trying to tell you what you should or shouldn't do. I've never been on your operations. That's for you to decide. I'm only sharing my story. But I'll tell you this much. At the end of '98, we were about as broke as broke can be. Now, within ten years I could retire and give this farm to my kids and they would never have to worry about money, and I won't, either. In ten years. The amount of profit we're making now is astronomical compared to before."

Gabe isn't afraid to appeal to the potential for higher profits. If he can prove that he's making more money with regenerative agriculture than conventional, then he has a farmer's ear to tell the rest of the story. Gabe talks about how much money they'll save by switching, explaining that it's tough to go broke under the regenerative model because most of the expenses farmers are worried about disappear, and the remaining costs, such as seed, hay, and fuel, shrink considerably. Again, he returns to his own experience. "This is a true story," he tells me. "I've had the same tax accountant for fifteen years now. I went to him last year and he says, 'Gabe, you've got problems. You have no expenses. You are going to end up paying a pile of taxes.' And I said, 'I know, and isn't that great?' That means I'm making money. So what? I don't mind paying some. But I'm not going to go out and buy a new tractor just to prevent paying taxes.[1] I won't do that. He says, 'But you got no expenses.' My chemical fertilizer is zero, no fungicides, no pesticides, my herbicide's a pittance, we don't use fuel because it's all no-till. I'm not using near the fuel we used to." Still, he hears objections. "So many people think, 'But I have equipment payments.' Well, do you have to have the equipment? Drop out what you don't need," Gabe says. I hear an echo of Kevin and his "for now we walk" philosophy. Gabe is down to a twenty-foot seed drill, three tractors (mostly for moving snow), a haying machine, a baler, a truck for hauling grain and seed, a grinder-mixer for processing feed for the poultry and hogs, and a few other small pieces of equipment. Everything but the drill is about twenty years old. Cutting out equipment payments, and the fuel and maintenance that go with the machines, has saved him hundreds of thousands of dollars.

For farmers who've already transitioned to regenerative agriculture, Gabe offers simple but revolutionary advice in the farming world: take your profit first. Here's how—after a few years of farming in the new model, Gabe says, a farmer knows roughly what he or she will produce in a year: a certain number of calves at about this weight, this many bushels of grain, or whatever the case may be. Gabe encourages producers to figure out what profit percentage they want (within reason), and then limit themselves to spending only what's left. That's not so hard in the regenerative model, he says, because expenses are so low. "It's virtually impossible when you think holistically to go broke," he says. "Obviously if you get hail four years in a row, it's possible, but that's highly unlikely. Nature can throw you a curveball, but then you just readjust."

"The thing of it is, you never go back and spend any of that profit that you were paid," he continues. "I tell producers, if you would have a job in town, your employer wouldn't be able to come and take those wages back from you. So why do we as farmers and ranchers allow our wages to be spent on inputs instead of putting it in our pockets?" This is why Gabe's advice is so revolutionary: farmers rarely put profits to off-farm use, if there are any. Under the conventional model, farmers and ranchers need to continually reinvest profit back into the operation so they can get bigger and avoid getting out. Just meeting conventional expenses is a challenge, let alone saving for expansion, so farmers tend to keep minimum living expenses for their families, but not much else.

This was a constant source of conflict between my parents while I was growing up. The farm soaked up money like a sponge, which rarely appeared to trouble my father but upset my mother, who was tasked with running a household of four children on a small budget. She felt that the family received mere drops of the revenue stream while the farm received a torrent. We always had enough to eat and wear and my parents left no bills unpaid, but by American middle-class standards, which I understand now that I live in a city, we lived very simply. Not that money and material possessions are life's most rewarding things, and obtaining them is not why Gabe, or even conventional farmers like my parents, choose to live on the land. Still, farms and ranches are businesses, not hobbies. They need to be financially viable. Gabe knows what it's like to be broke, and he doesn't want to go back there. He farms regeneratively mostly because of his principles, but partially because he would rather have a secure financial future. A cornerstone of his argument is that conventional farmers work harder for less money, while regenerative farmers work less and earn more. Convincing farmers to accept this fact is the first step in transitioning our nation's agriculture.

Another cornerstone is that the regenerative model offers farmers greater control. They are no longer forced to accept what the market will pay for commodities, the American farmer's struggle since the dawn of industrial agriculture. Instead, they can set their own prices or receive premiums for being organic, grass-fed, and so forth, which reduces stress. Gabe offers a concrete example. "I speak a lot out in the Corn Belt: Missouri, Indiana, Illinois, Ohio, all that," he says. "The average cost to produce a bushel of corn last year in those areas was right around $5 a bushel. Well, it cost me $1.42. Corn right now, I just saw it advertised in Minot, North Dakota, is $2.36. I can still make money. Not all that much, but I can still make money. There's not a lot of people who can at that." Under regenerative agriculture, farming would feel less like gambling and more like a thoughtful response to the land and consumers. Gabe recalls speaking to conventional farmers in Missouri the previous March. Everyone was chatting about how many acres of corn and soybeans they were going to plant and what price they were going to lock in on the futures market. Somebody asked Gabe what he planned to do. "I said, 'I pay no attention to the markets,'" he tells me. "Doesn't matter to me. For one, I've got so many different things I can market, and I'm controlling the price I receive. I set my own prices. So what do I care? The futures market and that, I don't have to pay any attention to that."

Like Phil, Gabe's independence from market forces has allowed him to do something most conventional farmers could never afford: he collects no government farm assistance to supplement his income. He's not alone. Many colleagues that go on speaking tours with him use a similar production model and have also opted out of government payment programs. "We're now out of all government programs," he says with pride. "I don't take direct payments, I don't take part in crop insurance. I was in an EQIP [Environmental Quality Incentives Program] contract.[2] And for one more year I still have a CSP contract; that's the Conservation Security Program. The only reason I didn't bow out of that is because it's more of a headache to NRCS [Natural Resources Conservation Service] if I do so because they would have a lot of paperwork to do, so they said, 'Just wait.'"

It's important to remember that not all crops are subsidized—most food crops like vegetables and fruits are not, and neither are livestock—but many commodity grain crops like corn, soy, wheat, and rice are. Most subsidies go to big farms: of the $20 billion doled out in 2005, the largest 10 percent of farms got 72 percent of the money, while 60 percent of farms received nothing.[3] Corn farmers pocketed almost half of that $20 billion.[4] But virtually all farmers and ranchers can purchase federally subsidized crop insurance and receive farm-related tax breaks, especially corporate farms and livestock operations. Federal and state programs prop up the industrial food system, a system no one could survive in without government help.

If we used government programs to support regenerative agriculture, then we would be able to justify their existence. Now, though, the system gives handouts to farmers and ranchers that are neither fair nor necessary, Gabe argues. "Does Bison have a café?" he asks me. Yes, I say. "Go to that café in Bison. Do they get their insurance premium subsidized? I'm sure not. So how come we do it for farmers? Is the café getting direct payments? No, but we give them to farmers. And I think that's wrong. Let's face it; this country has a huge debt problem. It's huge. So why should I, when I'm making plenty of money doing what I'm doing, take those? That's why I wanted to prove to people that it can be done. Our family just decided enough is enough. Somebody has gotta do it."

"I get really upset at farmers and ranchers and how they complain about people on welfare," he continues. I understand what he means: in general, farmers and ranchers tend to be politically conservative. They often view programs for the poor, like food stamps, assistance for single mothers, and unemployment benefits, as fiscally irresponsible and socially destructive because, they argue, these programs encourage laziness. Yet they don't see their own government payments as welfare. "I've actually said this when I'm out speaking to groups: I think every farmer or rancher should write the number down how much they paid the IRS, then start deducting your direct payments, your crop insurance subsidies, your EQIP contracts, your CSP contracts, and then ask yourself who's really on welfare. Of course in the Corn Belt they about come unglued. But it's true."

Ironically, conventional farmers have more in common with the poor than they realize. Government assistance aside, both farmers and the poor battle a system designed to keep them powerless. A combination of inadequate education, racism, classism, and many other factors often prevent the poor from securing decent-paying jobs and rising out of poverty. Instead of helping people up the socioeconomic ladder, America's social system perpetuates more poverty, much like the industrial food system and the treadmill of production push crop prices ever lower, encouraging consolidation, mechanization, and dependence on chemical agriculture. Farmers stuck on the treadmill of production are not much different than people trapped in poverty's vicious cycle. Still, politically conservative producers tend to argue that assistance to farmers is different because farmers are needed to feed the world—as if their human worth is somehow higher. In their minds, the farmer identity makes them more deserving of aid. Conventional farmers often fail to see how their "contributions" are actually making our world less habitable every day.

It's one thing for strangers to laugh at and reject you. It's another for your community to do that. For Gabe—and many farmers and ranchers who opt out of conventional agriculture—pushback comes from neighbors, friends, and family, the very people who are supposed to provide support. In places where conventional agriculture is the norm, like North Dakota, people tend to view nonconformists with suspicion and even hostility. "Every time I take another level and start doing one more thing, I see a lot of my old friends and . . ." Gabe's words trail off as he decides how to describe the criticism. "I'm still friends with them, but not near as close as we used to be, because they feel really uncomfortable with what I'm doing."

Practicing regenerative agriculture takes thick skin. While this requirement will likely go away as such agriculture becomes the norm, it's necessary for now. Regenerative farmers and ranchers have to be prepared for harsh criticism, social isolation, and arguments in person or in court. Think, for example, about a producer trying to grow organically when the neighbors next door spray their fields with Roundup, which runs off during rains, drifts over on the breeze, and creates virulent superweeds. If an organic field becomes contaminated, then the organic producer will lose his or her USDA organic certification and be forced to repeat the three-year waiting period as the land detoxifies. Meanwhile, conventional farmers, who are under extreme pressure to increase yields as quickly as possible, have little time to take extra precautions. GM crops pose a more serious contamination risk than agrochemicals—they've been known to cross with non-GM crops and appear without warning in other fields, ditches, or nonfarm areas. If cross breeding occurs, organic farmers can't save their seeds and must pay the GM seed company, such as Monsanto, a fee or face a lawsuit. It's not hard to see how neighbors could end up in court over issues like these.

Gabe has counseled farmers through the emotional stress that usually comes with transitioning to regenerative agriculture. He admits the transition is hard. "The one thing, without a doubt, that is the toughest for a lot of people to get used to is the scrutiny and criticism. That without a doubt is the hardest for a lot of people," he says. "For me it hasn't been much of a problem because, one thing is, I grew up in town. I'm not from a farm. People expected me to fail. They'll tell you that. I've had people walk up to me and say, 'Gabe, I never would have thought you would make it.' Though it's been a hindrance—I've had to learn everything the hard way—it's my greatest asset because I'm open-minded. There's nothing that I won't look at and consider." Being a "town kid," he faced criticism from the start, so farming regeneratively didn't change his social position like it does for conventional farmers. Still, his story reveals the double burden carried by farmers with nontraditional backgrounds. Whether they practice regenerative agriculture or not, people who move from urban areas to raise livestock or crops tend to experience some form of negativity from their neighbors. They might be labeled as naïve or written off as nothing but hobby farmers. Though Gabe has accepted the pushback, he's not afraid to push back, too. "One thing about me, I'm pretty outspoken. You're going to know where I stand," he says. Outspoken might be an understatement, I think to myself. "I'm not going to tell you you're stupid and a fool and all that. But I will challenge you and say, 'Why are you doing that?'"

"I've got many more people who want to learn," he continues. "They challenge me and help take me to the next level, and hopefully I do the same for them." I ask if he has any stories of people who've changed their operations after hearing him speak or going to one of his workshops. Gabe tells me about what he calls an unofficial fan club that developed in South Africa. A group of farmers followed his work for years online, then started growing cover crops and diversifying their livestock herds. Gabe had no idea this was happening until they emailed to ask if they could visit the farm, which they did. Similar groups exist in Australia, South America, and Canada. He offers another example: a week before I showed up, Gabe was in Missouri at the National Grassfed Exchange Conference, speaking to a crowd of three hundred or more people. During a discussion session, a young guy stood up and told the crowd that four years ago he didn't know where to go when it came to farming. He was stuck in the conventional model. Then he went to a meeting and listened to Gabe and the other speakers, and a light bulb turned on. Now the young man is using cover crops and mob grazing, and he's excited about farming again. Knowing his work has inspired others keeps Gabe motivated and helps him forget about the criticism he weathers both at home and out speaking. "You're like, wow, I am making a difference to somebody," Gabe says.

"It is real rewarding when we take our food to town, our meat and that, and you got people just falling all over you, so to speak, to get it, to buy it. That's rewarding, too," he continues. Some people might write off Gabe's feelings as just that: feelings, those subjective things that capitalism has little room for. Feeling rewarded by one's work doesn't pay the bills or generate profit, and it should rarely guide business decisions, the thinking goes. But Gabe's words resonate with me because they hint at something we rarely talk about when it comes to producing food: the psychological well-being of farmers and ranchers. Like other people, farmers and ranchers deserve and often crave emotional fulfillment from their labor. That rewarding feeling is something most of us strive for. As Viktor Frankl writes in *Man's Search for Meaning*, life offers three primary sources of meaning: creating a work or doing a deed, experiencing something or encountering someone, or choosing to find meaning in unavoidable suffering. Many Americans seek meaning in the first way, through work. No one says to the high school senior, "Seek a career that you find meaningless, one that provides no sense that you are contributing something good to the world." But we do not live in a perfect world, and therefore many people wind up working jobs that provide little to no fulfillment and, in too many cases, become depressed. That's what has happened to farmers. In the industrial agriculture model, they are simply cogs in the food-producing machine. They are worth little more consideration than the machines they operate. Everyone is encouraged to put production, profit, and technological progress ahead of the land and human health. No one is encouraged to seek personal fulfillment.

I think of my brother, who is already frustrated with conventional agriculture just four years in. I worry that if he doesn't change how the farm operates, he'll never find meaning in his work. I worry he will be controlled by the farm, as my father is, instead of rewarded and emotionally nourished by it. Even today, while Gabe and I are talking, Josh is at home on the swather, too busy haying to join me on the interview. He can't even take a day off for educational purposes or what the business world would call "continued training." I express my worries to Gabe and ask what advice he has for people who are looking to transition their farms and ranches to regenerative operations. What would you say to people who are overwhelmed by the scale of what needs to be done in order to start, as I suspect my brother is, or who are older and set in their conventional ways, like my father?

The first step is overcoming the fear of failure. Gabe says conventional farmers and ranchers worry about losing the control (or the illusion of control) that pesticides, herbicides, and synthetic fertilizers provide. Giving that control back to unpredictable nature can be scary. "When you work with nature, you've got to realize everything is in cycles," Gabe says. "It's not like you can have this uniform number of predators and prey, it's just a cycle. Some years some things are going to do better than others. That's been a challenge, learning how to accept that nature has those ebbs and flows, so to speak. You have to kind of learn to take what nature gives you."

Producers also need to stop making excuses for why regenerative agriculture won't work in their area. Like Gabe, I've heard a lot of these excuses over the years: the environment is too dry, too wet, too hilly, too hot, too cold, and so forth. Gabe says there is no such thing as an environment where holistic thinking doesn't apply. "It can be done anywhere because the principles are the same," he says. "I always hear, 'We don't get the moisture or this or that.' The principles are the same everywhere. There's nature everywhere. You're just mimicking nature is all you're doing." Despite the vast size and environmental differences between Phil's, Kevin's, and Gabe's operations, for example, they operate under the same basic principles, principles that could be applied to almost any farm, by any farmer. That's the only commonality we really need in our agriculture: a commitment to basic principles of regeneration.

The most important advice Gabe has for farmers and ranchers, though, is to keep learning. Seek new perspectives, read, be open to new ideas. "The biggest thing is, you have to have an open mind," Gabe says. "Then, find others who are doing it and are experienced. The one beautiful thing about this regenerative type of agriculture is, the majority of people doing it, they went through the transition themselves. A lot of my closest friends went through tough times, too, and that's what got them switched. When things are rosy, nobody's going to want to switch. I told people for the last several years, when corn prices were six-plus dollars, are you ready for when they are three? You better be getting ready now. If not now, when? So just start doing it."

"The other thing is, you gotta start slow," he continues. "Take baby steps. Like I tell people with cover crops, I say just try it on like forty acres, one field. But commit yourself. Say 'I'm going to do it for five years.' There's no going back. You either commit to that or not. Once they get through that, it's very few who go back."

Conclusion

Will my family make the change? As I'm writing this, my father remains firmly in the "It's too late for us" camp, having bought a new sprayer, another massive tractor, and a bigger seed drill. My brother, however, has started running his own small herd of grass-fed cattle using the rotational grazing model, and he planted two fields of cover crops. His goal is to regenerate native prairie much as Phil is doing and keep the cattle out of the CAFO system, and use diversified management on the farmland like Gabe does to restore the soil and wean his part of the farm off of inputs.

I am thrilled to see my brother taking these initial steps. Conventional farmers and ranchers, especially young ones like him, need to be included in the movement toward regenerative agriculture if any progress is to be made. Too often people resent the harmful consequences of conventional agriculture so much that they write these farmers off as part of the problem instead of the solution. The 2016 election exposed the many painful divides among us, and now more than ever we need to avoid an "us versus them" mentality. Those pushing for regenerative agriculture must welcome conventional producers to the table, listening to their concerns and being respectful of their needs. While it's true that industrial agriculture is an environmental, social, and economic tragedy, and many farmers and ranchers remain stubbornly resistant to change, it's also true that they are the people who do the actual work of farming. Consumers, politicians, and writers like me can talk about agriculture all we want, but we aren't on the land, making decisions that affect everyone. If we want change, then we have to partner with the folks on the land.

That's why people like Phil, Kevin, Fidel, Gabe, and their families give me hope. They show us not only that regenerative agriculture is possible, but also that farmers and ranchers can be stewards and still feed people and make good money. Their stories reveal that farmers get back far more than they give up by converting to regenerative agriculture. When producers use their gift of land ownership for good, the results benefit everyone as well as the natural world that nourishes all life. While the transition to regenerative production will likely be difficult, the good news is that we don't have to agree on a single route forward, because there isn't one. Just look at how different the farms profiled in this book are. Farmers and ranchers can choose the route that fits them and their land, working from a few broad principles: mimic nature, stack enterprises, revive soil, protect consumer health and the environment, be diverse, and use the sun's wealth. In a single phrase, treat the farm as a regenerative ecosystem.

If we do this, then we can drastically reduce our dependence on inputs like oil and agrochemicals. We can also avoid the fallacy of composition (the one-size-fits-all temptation), because every ecosystem will have unique needs and thus provide different crops and livestock. We can't all serve the same niche market or operate the same way, but we shouldn't try to anyway. That's when agriculture becomes vulnerable and top-heavy. Diversity will make our food system stronger and better equipped to handle crop failures, market collapses, and climate change. Our farms should have the diversity found in natural spaces, with a variety of plants, animals, and wildlife engaged in symbiotic relationships.

Farm country is full of good people with the resources to enact such change, but most of them are trapped in the conventional system. There comes a time, though, when claiming to be a good person trapped in a bad system is no longer an excuse for inaction—and that time is right now. We know that regenerative, organic agriculture can feed the world. We know that such agriculture is financially viable for families. And we know that conventional agriculture no longer works for our society or environment, if it ever did in the first place. People like Phil, Gabe, Fidel, and Kevin prove that farmers *can* opt out of the conventional system, and that even the most die-hard conventional farmer can start with forty acres of cover crops and go from there. To kick-start this transition, we need widely available training in regenerative agriculture. Such training could actually harness the stubborn streak in some farmers. Many value their independence, yet agribusiness corporations currently control them. Once they understand that regenerative agriculture allows far more financial and operational freedom than industrial agriculture, I think they will embrace the regenerative model.

Regenerative agriculture will be difficult to execute for some farmers and ranchers at first, mostly because it's so different than conventional production. Agriculture won't be as simple as zapping a field with Roundup or fattening livestock on corn. That's the beauty of regenerative agriculture, though: it's complex, just like nature. It puts the culture back in agriculture. It also restores the noble aspect of farming and ranching, making people stewards again instead of mere producers. Farmers needn't and shouldn't bear the burden of change alone, though. Consumers have to reengage with food production, educating themselves about regenerative agriculture and the artificial cheapness of conventional food. Only then will shoppers be willing to accept the true cost of food. We also need to reshape our food distribution system so that farmers and ranchers can feed their regions and communities. Then it's up to consumers to actually eat regionally and locally by supporting their farmers and farmers' markets, accepting that year-round availability of some items isn't sustainable. Consumers need to resist the urge to buy choice items only and instead eat from the whole farm, which will not only reduce waste, but also encourage farmers to grow responsibly and create a cuisine that celebrates what the land provides. Most of all, consumers need to value the work farmers and ranchers do and express that appreciation with their wallets as well as their votes.

Producers and consumers also need to work together to shape government policy in a way that supports regenerative agriculture. Laws alone won't cause change, but they can make it easier by incentivizing ecological practices, curtailing corporate influence over public policy and research, forcing producers to pay their workers fair wages, and turning existing institutions into resources for sustainability. Redirecting the mission of the Cooperative Extension Service and the land-grant colleges are just two of the many ways the federal government can do this. Strengthening the National Organic Program is another. If we want to achieve true sustainability, then we also have to redefine organic agriculture so that it isn't conventional agriculture in disguise. All of this is especially important now that we have climate change skeptics and pro-agribusiness leaders in cabinet positions following the 2016 election. The Trump administration clearly values corporate interests above the environment, social responsibility, and human health, so we have an uphill battle ahead. But I remain confident that common sense, science, and the will of the people will triumph in the end.

I believe this because such massive changes have been accomplished before. In one hundred or so years, we converted agriculture into a completely industrial act, such a short time in its ten-thousand-year history. Americans of the early twentieth century would have laughed at the idea of such a radical and all-encompassing transformation. It was only possible, though, because certain parties—corporations, agribusiness operators, large-scale farmers, and government officials—cared *a lot* about changing agriculture into an industrialized process. If a few players can create such extreme change in such a short period, imagine what a nation of united and passionate farmers and consumers can do. It's time for the people, urban and rural alike, to care a lot about changing agriculture, this time into a regenerative, life-sustaining act that works hand in hand with the natural world.

ACKNOWLEDGMENTS

This book would be nothing without Ryan Roth, Phil and Jill Jerde, Kevin O'Dare, Fidel Gonzalez, Gabe Brown, and all of their families. These folks welcomed me to their farms and, in some cases, into their homes, giving me entire days during the busiest seasons. They patiently answered my questions and explained their work with openness, good humor, and honesty. Each of them inspired my argument in different ways, and I am grateful for not only their help and time, but also their friendship.

I was blessed to have a community of talented writers and professors surrounding me during the writing of this book. Ayşe Papatya Bucak, Katherine Schmitt, and Andrew Furman read multiple drafts and offered insightful commentary, and they encouraged me on the road to publication. They also shared their creative wisdom in graduate writing workshops while I sought my MFA degree at Florida Atlantic University (FAU), and they remain valued mentors and friends. Thank you to the many friends and colleagues who read chapters of this book in workshops and gave thoughtful revision advice. My sincere thanks to FAU's English Department for providing the generous graduate assistantship that funded my MFA and generated this work, and for the Swann Award that helped support the writing of it. Thank you to FAU's College of Arts and Letters for granting two Advisory Board Graduate Student Awards that funded research-related travel. I would also like to thank Patrick Hicks, Janet Blank-Libra, and Jeffrey Miller for their meaningful contributions to my writing and thinking while I was an undergraduate student at Augustana University in Sioux Falls, South Dakota.

My agent, Matthew DiGangi at Bresnick Weil Literary Agency, steered me expertly through the publishing process by finding the right press, advocating for me tirelessly, and cheering me on. I am grateful for his expertise, insightful critiques, and unwavering optimism. I would also like to say thank you to the entire team at University of Nebraska Press for their excellent work in everything from cover design to publicity and beyond. I am especially grateful for my acquiring editor, Bridget Barry. She saw a spark in this manuscript and took the risk of signing a new writer, for which I am very thankful. Many thanks to copyeditor Karen Brown for an exceptionally thorough and perceptive manuscript review. This book was much improved by her attention to detail.

My parents, Leslie and Cathy Johnson, deserve the highest praise a child can give. They loved and encouraged me from the moment they brought me into this world. I am thankful for the beautiful childhood they gave me on the ranch, for their patience with me over the years, and for understanding why I needed to write this particular book (and for agreeing to appear in it, no matter what I wrote). My sisters, Anna Johnson and Charlotte Johnson, and brother, Joshua Johnson, are also my cheerleaders and very best friends, and I am grateful now and always for their love and reassurance.

Last but in no way least, I want to thank my husband, Ryan Anderson, who believed in me when I left a perfectly good career in communications to become a writer and teacher, who understood when I spent nights writing and weekends traveling, and who urged me on when the going got tough. He inspires and sharpens my work, and he wants me to be myself, no matter what. There is no one else I'd rather create a book or a life with.

NOTES

INTRODUCTION

1. Benyus, *Biomimicry*, 16.
2. Costello, *Prairie World*, 63.
3. Saunders et al., *Hotter and Drier*.
4. "Last Stand of the Tallgrass Prairie," National Park Service, July 18, 2017, http://www.nps.gov/tapr/index.htm.
5. Agribusiness has two definitions: it may be agriculture practiced using commercial standards, especially the application of advanced technology; and it may refer to a group of businesses dealing with agricultural produce and services to aid production. The "agribusiness" here means both, but primarily the latter. The powerful business players have developed a self-promotion system that has entangled supposedly objective media outlets such as *Tri-State Neighbor*.
6. I took a screen shot of Monsanto's website on May 17, 2017, at the URL below. At the time, the page included this quote: "We are proud to be a sustainable agriculture company." "Improving Agriculture," Monsanto, accessed May 17, 2017, http://www.monsanto.com/improvingagriculture/pages/default.aspx.

1. THE VICE PRESIDENT

1. About 444,000 acres of the EAA is sugarcane. Giant sugar corporations like Florida Crystals, King Ranch, and United States Sugar Corporation own most of those acres. Rocky Mountain National Park is 265,761 acres, if that helps put the EAA's sugarcane acreage in perspective. "Sugarcane, King of Crops," Palm Beach County History Online, accessed March 8, 2017, http://www.pbchistoryonline.org/page/sugarcane-king-of-the-crops.
2. An acre is 4,840 square yards, roughly the size of a football field.

3. During the vegetable and leaf part of the rotation, farmers usually harvest a crop and immediately plant a totally different one in its place—and repeat the process every few weeks or months depending on the crop's maturity cycle. A given block on Roth Farms is home to many different crops from September to April, whereas in places like Kansas or Illinois, where the growing season is short, one conventional field is home to one type of crop per summer.
4. Roth Farms also grows sweet corn and green beans in joint venture with other farms; Ryan helps a bit with land preparation on those crops, but that's it.
5. Hoppe and Banker, *Structure and Finances, 2010*, 2.
6. Hoppe, *Structure and Finances, 2014*, iii.
7. Data in this paragraph from Hoppe, *Structure and Finances, 2014*, iii–iv.
8. Data in this paragraph from Hoppe, *Structure and Finances, 2014*, iii–iv.
9. Koch and Peden, *Life and Selected Writings of Thomas Jefferson*, 377.
10. Hamilton, *Deeply Rooted*, 295.
11. Berry, *Unsettling of America*, 45–46.
12. Kramer, *Three Farms*, 197–274.

2. THE FARM WE GREW

1. Barber, *Third Plate*, 49.
2. Barber, *Third Plate*, 48–49.
3. Schlebecker, *Whereby We Thrive*, 25–35
4. I recognize that the story of the Plains Indians deserves far more attention than I can provide here. For further reading, start with Welch, *Killing Custer*, and Ewers, *Plains Indian History*. Later, Plains Indians started ranching and farming on the reservations and elsewhere, starting a unique cowboy culture that remains today.
5. Hurt, *Problems of Plenty*, 5.
6. Sulc and Tracy, "Integrated Crop-Livestock Systems," 335.
7. Sulc and Tracy, "Integrated Crop-Livestock Systems," 335.
8. Sulc and Tracy, "Integrated Crop-Livestock Systems," 335.
9. Hurt, *Problems of Plenty*, 13.
10. Pripps, *Big Book of Farmall Tractors*, 33.
11. It is interesting that these terms, industrial and agricultural, remain separate in 1930.
12. While looking into modern advertising rhetoric, I came across the referenced ad for the John Deere sprayer. I accessed the included quotes on December 3, 2014. The language has since been modified slightly as new sprayers have come on the

market. Deere & Company, "Self-Propelled Sprayers," accessed December 3, 2014, https://www.deere.com/en_US/products/equipment/self_propelled_sprayers/self_propelled_sprayers.page?.

13. For further reading on the history of American agriculture, see R. Douglas Hurt, *American Agriculture* and *The Problems of Plenty*; David Dary, *Cowboy Culture*; Barry Estabrook, *Tomatoland*; Michael Pollan, *The Omnivore's Dilemma*; Wendell Berry, *The Unsettling of America*; Eric Schlosser, *Fast Food Nation*; and Dan Barber, *The Third Plate*.
14. Information in this paragraph comes from Philpott, "Brief History."
15. Hurt, *Problems of Plenty*, 120.
16. Hurt, *Problems of Plenty*, 116.
17. Davis and Goldberg, *Concept of Agribusiness*, 2.
18. Pollan, *Omnivore's Dilemma*, 53.
19. Berry, *Unsettling of America*, 33.
20. Pollan, *Omnivore's Dilemma*, 51–52.
21. Very, very heavily: "Between 1970 and 1980, the amount of farm mortgage debt outstanding in the U.S. grew from $71.4 billion to $113.2 billion in constant 1982 dollars, an increase of 59 percent" (Barnett, "Farm Financial Crisis," 371).
22. An elevator is a grain storage tower that contains a lifting mechanism that scoops grain from the bottom. The term usually indicates the entire elevator complex—the offices, grain testing facility, storage units, and marketing service. Farmers sell their grain to the elevator instead of local buyers, which means selling it to a company that markets the grain into the food and livestock feed industries.
23. Pollan, *Omnivore's Dilemma*, 39.
24. WD refers to the Allis-Chalmers WD model tractor; M refers to the Farmall M model tractor. Both tractors were popular during the 1950s.
25. Barnett, "Farm Financial Crisis," 375–76.
26. Hurt, *Problems of Plenty*, 148–49.
27. Berry, *Unsettling of America*, 59.
28. Ikerd, *Crisis and Opportunity*, 4–5.
29. Schlosser, *Fast Food Nation*, 199–20.
30. Pollan, *Omnivore's Dilemma*, 54.
31. United States Department of Agriculture, *2012 Census of Agriculture*, volume 1, chapter 1: U.S. National Level Data, table 2, "Market Value of Agricultural Products Sold Including Landlord's Share and Direct Sales: 2012 and 2007," May

2014, http://www.agcensus.usda.gov/Publications/2012/Full_Report/Volume_1,_Chapter_1_US/.

32. P. Roberts, *End of Food*, 280.

3. THE GROWTH OF ROTH FARMS

1. "Muck" is the term for the soil exposed by draining wetlands, like the EAA's soil. Muck is deep black in color and made of organic matter in various states of decay—the floor of a former swamp.
2. Sugarcane is creating in Florida what corn has already been creating in the Midwest: a biological dead zone. On the 100,000-acre-plus sugarcane farms in the EAA, there is virtually no biodiversity. Growing the same crop over and over is stressful on the soil and encourages pests, weeds, and herbicide resistance. I find it compelling that a member of the grass family, sugarcane, has taken over the EAA, while a member of the same family, corn, has taken over most of the Midwest. Grass is poised to control more of South America, too, with the increasing number of sugarcane plantations and corn farms in Brazil, Argentina, Uruguay, and Paraguay.
3. Mealer, *Muck City*, 88.
4. Sarah Klein, Jacqlyn Witmer, Amanda Tian, and Caroline Smith DeWaal, "The Ten Riskiest Foods Regulated by the U.S. Food and Drug Administration," Center for Science in the Public Interest, October 6, 2009, http://www.cspinet.org/new/pdf/cspi_top_10_fda.pdf.
5. Madeline Drexler, "Foodborne Illness: Who Monitors Our Food?" The Schuster Institute for Investigative Journalism, September 16, 2011, http://www.brandeis.edu/investigate/foodborne-illness/who-monitors-food.html.
6. Albritton, *Let Them Eat Junk*, 83.

4. THE FARM TOWN

1. Nationally, eight of every ten thousand high school players make it to the NFL each year. From 1985 to 2012, Belle Glade has averaged about one player for every year. The school has just over one thousand students, meaning the town's representation in the NFL is astoundingly high, especially considering the poverty level. Mealer, *Muck City*, 11.
2. Ellyn Ferguson and Randy Loftis, "Plague Baffles Town: Belle Glade AIDS Rate Tops in U.S.," *Miami Herald*, August 11, 1985, PDF, 1.

3. "Uniform Crime Report, 2003," Federal Bureau of Investigation, October 27, 2004, http://www.fbi.gov/about-us/cjis/ucr/crime-in-the-u.s/2003.
4. Lisa Rab, "Belle Glade Faces its Demons after a Senseless Murder," *Broward/Palm Beach New Times*, March 22, 2012, http://www.browardpalmbeach.com/2012-03-22/news/belle-glade-faces-its-demons-after-a-senseless-murder/full/.
5. Big Sugar, like Big Corn, is propped up by the federal government with price supports, domestic marketing allotments (American sugar growers are guaranteed 85 percent of the market, no matter what), and tariffs on imported sugar. So the government encouraged sugar growers to expand by crafting favorable policies, a move that damaged local economies and created an "empire of sugar" that holds too much power, like Big Tobacco and Big Oil.
6. Ikerd, *Crisis and Opportunity*, 5.
7. Information about tomato pickers, the piece system, and overtime pay is from Estabrook, *Tomatoland*, xix.
8. Rodriguez, "Cheap Food," 125–30.
9. Literature has often focused on impoverished farm workers. The plight of workers in the Chicago stockyards is chronicled in the fiction-but-practically-nonfiction book *The Jungle* by Upton Sinclair. John Steinbeck's famous novel *The Grapes of Wrath* records the horrifying story of an Oklahoma farm family that fled the Dust Bowl, only to find themselves starving and broke working on California farms.
10. Information about the *bracero* program is from Hurt, *Problems of Plenty*, 102.
11. Estabrook, *Tomatoland*, xix.
12. Estabrook, *Tomatoland*, xvii.
13. Rodriguez, "Cheap Food," 125.

5. THE MUCK

1. Technically, the EAA has two main soil types, muck and sand. The muck gives way to sandy, mineralized soils at the edges of the EAA, closer to the ocean. Muck is more prevalent, however.
2. M. R. McDonald, "Management of Organic Soils," Ontario Ministry of Agriculture, Food, and Rural Affairs, AgDex No. 510, last modified April 2010, http://www.omafra.gov.on.ca/english/crops/facts/93-053.htm.
3. Muck facts are from the University of Idaho's "Twelve Soil Orders: Soil Taxonomy," University of Idaho College of Agricultural and Life Sciences, last modified November 2013, http://www.cals.uidaho.edu/soilorders/; and from George Silva, "Keeping Muck Soils Sustainable," Michigan State University

Extension, November 2, 2012, http://msue.anr.msu.edu/news/keeping_muck_soils_sustainable.

4. Information in this paragraph is from Wright and Hanlon, "Organic Matter," 1–4.
5. Ingebritsen et al., "Florida Everglades," 101.
6. "Did You Know? Facts about EREC," Everglades Research and Education Center, University of Florida, July 25, 2011, http://erec.ifas.ufl.edu/did_you_know_facts.shtml.
7. Tihansky, "Sinkholes, West-Central Florida," 2–3.
8. Muck information in this paragraph comes from Ingebritsen et al., "Florida Everglades," 95–106.
9. Information on Iowa soil is from Wilde, "Study: Soil Eroding Faster than Estimated."
10. A collection of five pillars at a rest stop along Highway 80 in Adair County, Iowa, illustrates this loss. The first pillar, representing the soil level in 1850, is the tallest. The pillars grow steadily shorter until the year 2000 (the artwork was installed in 2002). Black stems of grass decorate the upper portion of each pillar to commemorate the tallgrass prairie Iowa once was. Kind of spooky how art can make something as devastating as soil loss look stately.
11. Ingebritsen et al., "Florida Everglades," 103.
12. EAA farming practices and subsidence data are from Wright and Hanlon, "Organic Matter," 1–4.

6. THE HOLISTIC PHILOSOPHY

1. United States Department of Agriculture, *2012 Census of Agriculture*, volume 1, chapter 1: U.S. National Level Data, table 34, "Other Animals and Animal Products – Inventory: 2012 and 2007," May 2014, http://www.agcensus.usda.gov/Publications/2012/Full_Report/Volume_1,_Chapter_1_US/.
2. In 2009, artist and technologist Stephen Von Worley mapped the McDonald's in the lower forty-eight states at the time, revealing that the McFarthest Spot lies between Glad Valley and Meadow, just east of my parents' house. The McFarthest Spot has since moved to a lonely spot in Nevada. Stephen Von Worley, "Where the Buffalo Roamed: How Far Can You Get from McDonald's?" *Data Pointed*, September 22, 2009, http://www.datapointed.net/2009/09/distance-to-nearest-mcdonalds/.
3. We routinely name our vehicles. There's also New Blue, a 1999 diesel that Dad bought in 2014, hence the "new" part. There's also Gary, a '90s-era hand-me-down from my dad's father. Its previous owners applied printed decals of their names

by each door: Gary and Jean. Those are not my grandparents' names, but they kept the decals on the entire time they owned the truck, as my dad still does.

4. Saunders et al., "Hotter and Drier," 2.
5. Savory, *Holistic Management*, 17–19.
6. Pollan, *Omnivore's Dilemma*, 213.
7. I think the following is a particularly insightful part of the chapter: "Instead of fixing what's really broken or finding a fundamentally different path, we print more money, invent a new drug, make a bigger bomb, suppress or buy off dissent, or build a dam . . . think carefully about what might be causing your problem." Savory, *Holistic Management*, 274.
8. As an example, Savory encourages all people to consider the effects of their daily actions: "Each day we put the utmost concentration and energy into our chosen tasks, seldom reflecting that we work within a greater whole that our actions will affect, slowly, cumulatively, and often dramatically. In our culture it is mainly philosophers who concern themselves with this larger issue because it is hard to see how individuals caught up in daily life can take responsibility for the long-term consequence of their actions, but they can. We can" (Savory, *Holistic Management*, 17).
9. Matthew Cawood, "More Livestock Is Climate Change Key," *The Land*, July 10, 2011, http://www.theland.com.au/news/agriculture/cattle/general-news/more-livestock-is-climate-change-key/2218226.aspx?storypage=0.
10. "Time Line of the American Bison," U.S. Fish and Wildlife Service, accessed September 6, 2014, http://www.fws.gov/bisonrange/timeline.htm.
11. "Time Line of the American Bison," U.S. Fish and Wildlife Service, accessed September 6, 2014, http://www.fws.go/bisonrange/timeline.htm.
12. Allred, Fuhlendorf, and Hamilton, "The Role of Herbivores in Great Plains Conservation," 1.
13. Ranchers commonly refer to their livestock as "critters." Though the term sounds strange, it's not derogatory. It's like cowboys of old calling the cattle "little doggies" on trail drives.

7. THE GRASS

1. Savory, *Holistic Management*, 38.
2. That buffalo doesn't live on the same fifteen acres year-round. The stocking rate is a simple measure of how many livestock a ranch can support, in this case the ranch's grazing acres divided by the number of buffalo.

3. D'Odorico et al., "Global Desertification," 328.
4. "Rural Poverty and Desertification," International Fund for Agricultural Development, August 2010, https://www.ifad.org/documents/10180/77105e91-6f72-44ff-aa87-eedb57d730ba.
5. Karl, Melillo, and Peterson, eds., *Global Climate Change Impacts*, 83–84.
6. Saunders et al., "Hotter and Drier," v.
7. Saunders et al., "Hotter and Drier," 29–30.
8. Saunders et al., "Hotter and Drier," iv.
9. Vose et al., "Temperature Changes in the United States," 185–206.
10. Barber, *Third Plate*, 70–71.
11. Ranchers have been told this by university researchers, Cooperative Extension Service educators, and other grazing "experts" like seed companies. Many ranchers believe cool-season grasses are the norm. This is another example of farmers and ranchers being told how to operate by agribusiness, to their detriment.
12. Costello, *Prairie World*, 65.
13. Tober and Chamrad, "Warm-Season Grasses," 227.
14. Information about grass types in this paragraph is from Tober and Chamrad, "Warm-Season Grasses," 227.
15. Ivomec is a commonly used liquid parasiticide that ranchers pour on their cattle to kill roundworms, lungworms, grubs, sucking lice, biting lice, mange mites, and horn flies.
16. Bowman and Zilberman, "Economic Factors," 34.
17. Boxler, "Fly Control for Cattle on Pasture in Nebraska," University of Nebraska–Lincoln Beef Division, May 2016, https://beef.unl.edu/cattleproduction/controllingflies.
18. Burns, Collins, and Smith, "Plant Community Response to Loss of Large Herbivores," 2337–39.
19. Burns, Collins, and Smith, "Plant Community Response to Loss of Large Herbivores," 2329.

8. THE BUFFALO

1. USDA National Nutrient Database for Standard Reference, accessed July 12, 2013, http://ndb.nal.usda.gov.
2. Barber, *Third Plate*, 114–15.
3. Barber, *Third Plate*, 115.

4. "Summary of Important Health Benefits of Grass-fed Meats, Eggs, and Dairy," Eat Wild, accessed July 12, 2013, http://www.eatwild.com/healthbenefits.htm.
5. Brucellosis is a livestock disease of European origin that devastated U.S. herds in the early twentieth century and also spread to elk and wild bison. Since 1934 the Cooperative State Federal Brucellosis Eradication Program has encouraged ranchers to vaccinate for this disease that's known as undulating fever or Malta fever in humans. Few buyers will purchase livestock that are not Bangs vaccinated.
6. Weaning is a natural occurrence, heartbreaking as it is to watch. Mammals do not provide milk for their young indefinitely. After half a year or more, a cow's body is ready to stop giving milk to feed a five-hundred-pound "baby." She's often pregnant with next year's calf as well, and sometimes she's already weaned the calf on her own by this point. Despite the calf's two days of bawling for his mother, the weaning would have happened anyway.
7. Grain is not inherently bad. For hundreds of years, farmers have fed livestock grain for the month or so leading up to slaughter. The ratios in CAFOs, however, are far too grain-heavy, last too long, and include the toxic grain from conventional farms.
8. Albritton, *Let Them Eat Junk*, 102–3.
9. Albritton, *Let Them Eat Junk*, 102.
10. Remember how Savory pointed out that herbivores don't like to stay long on areas they've soiled, which is why they move often? It must be torture for our steer to stand, day in and day out, in his own waste along with the waste of thousands of other cattle.
11. Sabrina Tavernise, "Farm Use of Antibiotics Defies Scrutiny," *New York Times*, September 3, 2012, http://www.nytimes.com/2012/09/04/health/use-of-antibiotics-in-animals-raised-for-food-defies-scrutiny.html?pagewanted=all&_r=0.
12. Raloff, "Hormones," 11.
13. "Interview with Michael Pollan," PBS Frontline, June 30, 2002, http://www.pbs.org/wgbh/pages/frontline/shows/meat/interviews/pollan.html.
14. Hribar, *Understanding Concentrated Animal Feeding Operations*, 5.
15. Sabrina Tavernise, "Antibiotics in Livestock: FDA Finds Use Is Rising," *New York Times*, October 2, 2014.
16. Albritton, *Let Them Eat Junk*, 111.

17. "Interview with Michael Pollan," PBS Frontline, June 30, 2002, http://www.pbs.org/wgbh/pages/frontline/shows/meat/interviews/pollan.html.
18. "Interview with Michael Pollan," PBS Frontline, June 30, 2002, http://www.pbs.org/wgbh/pages/frontline/shows/meat/interviews/pollan.html.
19. "Bison by the Numbers," National Bison Association, https://bisoncentral.com/bison-by-the-numbers/.

9. THE END OF THE CAFO

1. USDA Livestock Slaughter 2016 Summary, ISSN: 0499–0544, National Agriculture Statistics Service, April 2017, http://usda.mannlib.cornell.edu/usda/current/LiveSlauSu/LiveSlauSu-04-19-2017.pdf, p. 6.
2. USDA Livestock Slaughter 2016 Summary, 6.
3. "Investor Fact Book – Fiscal Year 2017," Tyson Foods, Inc., 2017, PDF, 4, http://ir.tyson.com/investor-relations/investor-overview/tyson-factbook/.
4. See Schlosser, *Fast Food Nation*, for further information about labor conditions inside slaughterhouses.
5. "Interview with Michael Pollan," PBS Frontline, June 30, 2002, http://www.pbs.org/wgbh/pages/frontline/shows/meat/interviews/pollan.html.
6. Lim et al. "Brief Overview of *Escherichia coli*," 5–14.
7. Information in this paragraph is from "Interview with Michael Pollan," PBS Frontline, June 30, 2002, http://www.pbs.org/wgbh/pages/frontline/shows/meat/interviews/pollan.html.
8. Reed, *Rebels for the Soil*, 107.
9. Reed, *Rebels for the Soil*, 99–100.
10. The Organic Watergate—White Paper. "Connecting the Dots: Corporate Influence at the USDA's National Organic Program" (Cornucopia WI: Cornucopia Institute, May 2012), 3, PDF.
11. The Organic Watergate, "Connecting the Dots," 4.
12. The Organic Watergate, "Connecting the Dots," 4.
13. Take a look at an infographic called "Who Owns Organic" here: http://www.cornucopia.org/wp-content/uploads/2014/02/Organic-chart-feb-2014.jpg. It's astounding.
14. Carey Gillam, "After Washington GMO Label Battle, Both Sides Eye National Fight," Reuters, November 8, 2013, http://www.reuters.com/article/2013/11/08/us-usa-gmo-labeling-idusbre9a70uu20131108.

15. Information in the preceding paragraphs about organic beef production and slaughter is from "Organic Livestock Requirements," National Organic Program, Agricultural Marketing Service, USDA, February 2013, https://www.ams.usda.gov/sites/default/files/media/Organic%20Livestock%20Requirements.pdf; and "Organic Meat and Poultry Processing Basics," Minnesota Department of Agriculture, March 2005, http://www.mda.state.mn.us/Global/MDADocs/food/organic/organicmeatprod.aspx.
16. Albritton, *Let Them Eat Junk*, 155.

10. THE SUN'S WEALTH

1. Gladwell, *Tipping Point*, 179–87.
2. Albritton, *Let Them Eat Junk*, 186.
3. Albritton, *Let Them Eat Junk*, 193.
4. Albritton, *Let Them Eat Junk*, 186.
5. Kenner, "Exploring the Corporate Powers Behind the Way We Eat," 37.
6. Cain, "Food, Inglorious Food," 275.
7. Albritton, *Let Them Eat Junk*, 189.
8. "2012 Presidential Race," Center for Responsive Politics, accessed July 2, 2013, https://www.opensecrets.org/pres12/. To put these numbers into context, super PAC contributions totaled $200 million for Romney and $58 million for Obama. John Hudson, "The Most Expensive Election in History by the Numbers," *Atlantic Wire*, November 6, 2012, http://www.thewire.com/politics/2012/11/most-expensive-election-history-numbers/58745/.
9. "Top Industries, Federal Election Data for Hillary Clinton, 2016 Cycle," Center for Responsive Politics, accessed May 16, 2017, https://www.opensecrets.org/pres16/industries?id=N00000019; and "Top Industries, Federal Election Data for Donald Trump, 2016 Cycle," Center for Responsive Politics, accessed May 16, 2017, https://www.opensecrets.org/pres16/industries?id=N00023864.
10. Adam Liptak, "Justices, 5–4, Reject Corporate Spending Limit," *New York Times*, January 21, 2010, http://www.nytimes.com/2010/01/22/us/politics/22scotus.html?pagewanted=all&_r=0.
11. Schlosser, *Fast Food Nation*, 261.
12. Schlosser, *Fast Food Nation*, 273–75.
13. The average corn yield in the U.S. is 158.8 bushels per acre. In Iowa, some fields yield well over 200 bushels an acre.

14. Liquid anhydrous ammonia is one form of synthetic nitrogen. Farmers purchase it by the tank-load and apply it with special injection drills. This form of nitrogen fertilizer can burn the skin severely and can cause death by asphyxiation. As Extension Specialist John Shutske warns farmers, "Anhydrous ammonia is caustic and causes severe chemical burns. Body tissues that contain a high percentage of water, such as the eyes, skin, and respiratory tract, are very easily burned. Victims exposed to even small amounts of ammonia require immediate treatment with large quantities of water to minimize the damage." He urges farmers to wear protective goggles, rubber gloves, and heavy-duty, long-sleeved shirts. Anyone who stores anhydrous in bulk should carry a rainsuit and two gas masks. This is what our food plants "eat," as do we when we consume them. John Shutske, "Using Anhydrous Ammonia Safely on the Farm," University of Minnesota Extension Service, WW–02326, 2005, http://www.extension.umn.edu/agriculture/nutrient-management/nitrogen/using-anhydrous-ammonia-safely-on-the-farm/.
15. Albritton, *Let Them Eat Junk*, 151.
16. Philpott, "Brief History."
17. Pollan, *Omnivore's Dilemma*, 42.
18. Pollan, *Omnivore's Dilemma*, 44.
19. Pollan, *Omnivore's Dilemma*, 45.
20. Pollan, *Omnivore's Dilemma*, 46.
21. Albritton, *Let Them Eat Junk*, 22.

11. THE SURFING FARMER

1. At first I thought it strange for a farmer to live half a block from the beach in a neighborhood chock-full of fabulously wealthy people, but Kevin is truly a surfer at heart and he couldn't bear to live too far from the ocean. He and his wife purchased the house thirty-five years ago, when beachfront houses were more affordable than they are today. Plus, his farm is too small to include housing.
2. Barber, *Third Plate*, 87.
3. Barber, *Third Plate*, 94.
4. Barber, *Third Plate*, 97–98.
5. Weeds are generally understood as the farmer's mortal enemy, which makes Kevin's moderate view surprising. The multibillion-dollar herbicide industry reinforces the view that weeds are evil plants. Organic farmers use other weed-control measures in addition to plastic, such as crop rotation, mechanical tillage, hand-weeding, cover crops, mulches, and flame weeding.

6. These strategies and others described later work for regenerative farming in general, not organic regenerative only.
7. "The Farming Systems Trial: Celebrating 30 Years," Rodale Institute, October 2011, http://rodaleinstitute.org/assets/FSTbookletFINAL.pdf, 8.

12. THE MISSION

1. Wilcher, "Greening of Milton Criticism," 1021.
2. Wilcher, "Greening of Milton Criticism," 1021.
3. Wilcher, "Greening of Milton Criticism," 1021.
4. Leopold, *Sand County Almanac*, viii.
5. For more information on organic standards, see "Organic Production and Handling Standards," National Organic Program, Agricultural Marketing Service, U.S. Department of Agriculture, November 2016, https://www.ams.usda.gov/publications/content/organic-production-handling-standards.

13. THE PLANTS

1. "Acreage," USDA National Agriculture Statistics Service, June 2017, 29 and 31, http://usda.mannlib.cornell.edu/usda/current/Acre/Acre-06-30-2017.pdf.
2. Monsanto once manufactured DDT, Agent Orange, and PCBs, and it has a dark history that includes releasing toxins into a river and giving an entire community cancer, suing farmers and journalists, and much more. For more on Monsanto, see *Lords of the Harvest: Biotech, Big Money, and the Future of Food* by Daniel Charles and *The World According to Monsanto: Pollution, Corruption, and the Control of the World's Food Supply* by Marie-Monique Robin.
3. United States Department of Agriculture, *2012 Census of Agriculture*, "Highlights: Farm and Farmland," September 2014, https://www.agcensus.usda.gov/Publications/2012/Online_Resources/Highlights/Farms_and_Farmland/Highlights_Farms_and_Farmland.pdf.
4. National Research Council, *Impact of Genetically Engineered Crops*, 4.
5. Philpott, "Monsanto GM Soy."
6. Conca, "It's Final."
7. Conca, "It's Final."
8. Pollan, *Omnivore's Dilemma*, 40.
9. Conca, "It's Final."
10. Gassmann et al., "Field-Evolved Resistance to Bt Maize," 1.
11. Johnson and O'Connor, "These Charts Show Every Genetically Modified Food."

12. Information about Seralini is from the 2013 film *GMO OMG*, directed by Jeremy Seifert, and "Controversial Seralini Study Linking GM to Cancer in Rats Is Republished," *The Guardian*, June 24, 2014, http://www.theguardian.com/environment/2014/jun/24/controversial-seralini-study-gm-cancer-rats-republished.
13. "The Farming Systems Trial: Celebrating 30 Years," Rodale Institute, October 2011, http://rodaleinstitute.org/assets/FSTbookletFINAL.pdf.
14. Let's not take the Rodale Institute's word alone. A sixteen-year (and ongoing) study, called the Long-Term Agroecological Research (LTAR) experiment, conducted by Iowa State University's Leopold Center for Sustainable Agriculture that compares organic and conventional crop rotations experienced results similar to Rodale's. See https://www.leopold.iastate.edu/long-term-agroecological-research for more information.

14. THE LIFESTYLE

1. "Platinum Clubs of America 2016-18: Top 150 Country Clubs," John Sibbald Associates, Inc., 2016, https://clubleadersforum.com/pcoa/the-2016-lists/top-150-country-clubs-2016/.
2. Apparently the fax machine has not gone out of vogue with some Vero chefs. As Kevin says, "In the middle of the night I hear my fax machine go off and I'm like, money!"
3. Berry, *Unsettling of America*, 37–38.

15. THE CONSUMER

1. Obach, "Theoretical Interpretations," 229.
2. Obach, "Theoretical Interpretations," 232.
3. Obach, "Theoretical Interpretations," 230.
4. Obach, "Theoretical Interpretations," 234.
5. Albritton, *Let Them Eat Junk*, 180.
6. "The Emissions Gap Report 2013," United Nations Environmental Programme (UNEP), November 2013, 24, http://web.unep.org/sites/default/files/EGR2013/EmissionsGapReport_2013_high-res.pdf.
7. By no means do I think this is the case with every farming community or every farmer. There are those who want to change, but do not know how to or are afraid to. I base my comments on rural communities here and throughout on eighteen

years of living in a farming community and my year with *Tri-State Neighbor*, plus the interviews for this book.

8. *The Emissions Gap Report 2013*, United Nations Environmental Programme (UNEP), November 2013, 24, http://web.unep.org/sites/default/files/EGR2013/EmissionsGapReport_2013_high-res.pdf.
9. "Extension," USDA National Institute of Food and Agriculture, accessed April 30, 2018, https://nifa.usda.gov/extension.
10. Berry, *Unsettling of America*, 151.
11. Berry, *Unsettling of America*, 151–52.
12. A contentious goal, because reducing agriculture to a science is arguably impossible: "Because the soil is alive, various, intricate, and because its processes yield more readily to imitation than to analysis, more readily to care than to coercion, agriculture can never be an exact science. There is an inescapable kinship between farming and art, for farming depends as much on character, devotion, imagination, and the sense of structure, as on knowledge. It is a practical art." Berry, *Unsettling of America*, 87.
13. Information in this paragraph is from Food & Water Watch, *Public Research, Private Gain: Corporate Influence Over University Agricultural Research*, April 26, 2012, https://www.foodandwaterwatch.org/news/public-research-private-gain-corporate-influence-over-university-agricultural-research.
14. Food & Water Watch, *Public Research, Private Gain*, 15.

16. THE FARMER GOES TO THE TABLE

1. *Occupational Outlook Handbook, 2014–15 Edition*, Bureau of Labor Statistics, January 2014, http://www.bls.gov/ooh/.
2. Dimitri and Oberholtzer, "Marketing U.S. Organic Foods," iii.
3. Dimitri and Oberholtzer, "Marketing U.S. Organic Foods," 4–5.
4. Dimitri and Oberholtzer, "Marketing U.S. Organic Foods," 3.
5. Joe Satran, "Organic Agriculture Benefits Revealed in New Long-Term Study from Rodale Institute," *Huffington Post*, October 5, 2011, http://www.huffingtonpost.com/2011/10/06/organic-agriculture-benefits_n_998214.html.
6. Information in this paragraph is from Charlotte Vallaeys, "Busting the 'Organic is Expensive' Myth," Cornucopia Institute, October 23, 2014, http://www.cornucopia.org/2013/10/busting-organic-expensive-myth/.
7. Rodriguez, "Cheap Food," 128.

8. Not always, as we saw in Phil's case. His sustainable system requires very little work because he doesn't spend time haying, administering parasiticides, or feeding during the winter.
9. See the *Washington Post*'s interactive map for searching zip code data here: http://www.washingtonpost.com/sf/local/2013/11/09/washington-a-world-apart/.
10. Fred Kirschenmann, Steve Stevenson, Fred Buttel, Tom Lyson and Mike Duffy, "Why Worry about the Agriculture of the Middle?" *Leopold Center Pubs and Papers* 143, 2004, https://lib.dr.iastate.edu/Leopold_pubspapers/, 143.

17. THE URBAN FARMER

1. The AFSC is a Quaker organization that implements service, development, and peace programs around the world. The New Mexico branch has other programs in addition to the farmer-to-farmer training program.
2. Bustos is a legend in New Mexico's organic and sustainable farming movement. He operates a certified organic vegan family farm, one of only a few in the nation, and uses indigenous growing techniques. In January 2014, he was one of a small number of elder organic farmers invited to California for a special conference on the progression of organic farming (the event appeared in the *New York Times* and the *Washington Post.*) He was the New Mexico State University Leyendecker Agriculturist of Distinction in 2005, the New Mexico Farmer of the Year in 2006, and the New Mexico Organic Commodity Commission organic farmer of the year in 2011.
3. Wozniak, *Irrigation in the Rio Grande Valley*, 10. Wozniak notes that some scholars maintain that New Mexican irrigation systems appeared well before 1500, but he's skeptical of these claims. The bottom line is that ditch irrigation has been practiced in New Mexico for a long time, and today it's considered a form of traditional agriculture.
4. Information about New Mexico's current drought and its dust storms is from Laura Paskus, "Muddy Hymnal: New Mexico's Drying Rivers Herald a Changing World," *Santa Fe Reporter*, June 18, 2013, http://www.sfreporter.com/santafe/article-7495-muddy-hymnal.html.
5. Ault et al., "Assessing the Risk of Persistent Drought," 7529.
6. CSA programs work like this: families purchase shares of the farm's produce in advance (one box a week, two boxes, etc.), with the farmer determining how many shares he or she can sell each week depending on the size and production of the farm. Some CSA programs deliver the shares to the families; with others,

the families go to the farm and pick up the shares. The CSA program I was part of left weekly shares at my neighborhood Whole Foods, where I picked them up.

7. The national average was 14.5 percent in 2013, per the U.S. Census Bureau.
8. "Demographics," Albuquerque Economic Development, accessed October 5, 2014, http://www.abq.org/Demographics.aspx.

18. THE AGRICULTURALIZED CITY

1. Berry, *Unsettling of America*, 90.
2. "South Valley," La Entrada Realty, accessed April 30, 2018, http://www.laentradarealty.com/albuquerque-real-estate-areas/south-valley.
3. Land Values: 2014 Summary, USDA National Agricultural Statistics Service, ISSN: 1949–1867, August 2014, http://usda.mannlib.cornell.edu/usda/nass/AgriLandVa/2010s/2014/AgriLandVa-08-01-2014.pdf, p. 6.
4. Climate Action Plan, City of Albuquerque website, accessed May 12, 2017, https://www.cabq.gov/cap.
5. The others are business, industry and carbon offset; carbon-neutral buildings; clean, renewable energy; complete, livable neighborhoods; recycling and zero waste; social change; and transportation.

19. THE DIVERSIFIED FARM

1. A word on no-till: No-till farming was developed in the 1960s and is an increasingly used approach to tillage worldwide or, more specifically, the lack thereof. Farmers use specialized seeders to sow through the residue with minimum soil disruption. No-till reduces erosion and increases soil fertility, and it was intended to conserve topsoil that was being rapidly depleted by conventional farming. Those benefits haven't gone away, but the caveat is that today's version of no-till is better termed "no-till chemical farming." Instead of plowing weeds under, farmers use herbicides to control them, which contributes to weed resistance, agrochemical runoff, and higher input costs. Gabe rarely uses herbicide, though, so his execution of no-till farming qualifies as regenerative. Huggins and Reganold, "No-Till," 71–77.
2. Today Brown's Ranch would be able to weather a one-year drought without a problem—the soils are so healthy and store so much water that even in dry years, Gabe still produces high-yielding crops and pasture. But in 1997, the soils were still recovering from decades of heavy tillage.
3. Before North Dakota was settled and plowed up, the soil contained approximately 7 percent to 8 percent organic matter, Gabe tells me.

20. THE SOIL

1. The soil information in this paragraph is from Wolfe, *Tales from the Underground*, 93–94.
2. Wolfe, *Tales from the Underground*, 94–95.
3. Barber, *Third Plate*, 86–87.
4. Wolfe, *Tales from the Underground*, 1.
5. Herbicide kills weeds, whereas fungicides kill plant diseases and pesticides kill bugs. While it's not an organic practice, Gabe's use of herbicide every two to three years is pretty impressive because he does not till to control weeds like many organic farmers do. Most conventional corn and soybean farmers spray for weeds before planting, again during the growing season if necessary, and again after harvest.
6. North Dakota is part of America's drylands, which encompass most of the Great Plains and the West. Drylands are areas in which the growing season is 1 to 179 days, which includes regions classified as arid, semiarid, and dry subhumid. About 40 percent of the world's land is classified as drylands. Koohafkan and Stewart, *Water and Cereals in Drylands*, 5–6.
7. Another form of multicropping is strip cropping, or planting long strips of different crops side by side, but in wide enough strips to permit machine harvesting.
8. Barber, *Third Plate*, 65.
9. Some people would argue that the things I've listed as intangible gains are in fact tangible. One can see water infiltrating better, for example. These kinds of benefits are not connected to a monetary value, though, which is why I call them intangible. If industrial farmers can't see the money made from covers on a balance sheet, then the benefits tend to be intangible and therefore unimportant to them.
10. Barber, *Third Plate*, 14–15, 430–31.
11. Root exudates are the chemicals exuded by plant roots into the soil. Scientists understand little about roots in general and even less about root exudates, but it is believed that the secreted chemicals help plants regulate the soil microbe community in their immediate vicinity, cope with herbivores, encourage beneficial symbioses, change the chemical and physical properties of the soil, and inhibit the growth of competing plant species. Soil microorganisms consume root exudates, as Gabe points out here when discussing the performance of soil biology in monocultures and polycultures. Walker et al., "Root Exudation and Rhizosphere Biology," 44.

12. Many of the plants have multiple benefits to the environment. The fusilli, for instance, also helps increase organic matter in the soil. "It has a tremendous root mass," Gabe says. "Two-thirds of your organic matter increase will come from roots; that's why we put species like this in here."
13. Jonathan Lundgren, "Tempering the Bad Reputation of Earth's Most Abundant Animals," *Tri-State Neighbor*, April 21, 2013, http://www.tristateneighbor.com/news/regional/tempering-the-bad-reputation-of-earth-s-most-abundant-animals/article_d62feb22-a85f-11e2-8934-001a4bcf887a.html.
14. Jonathan Lundgren, "Tempering the Bad Reputation."
15. Buhler, "Introduction to Insecticide Resistance."

21. THE ABUNDANCE OF AN ACRE

1. Pollan, *Omnivore's Dilemma*, 215.
2. A good yield for wheat in eastern Colorado's dryland climate would be around sixty bushels per acre.
3. A quarter of land is 160 acres. I find it ironic that 160 acres, a quarter, was the exact amount given to homesteaders under the Homestead Act of 1862. Maybe the government wasn't too far off in its estimation that 160 acres could support a family. Of course, when Gabe says 160 could support a family, he means if it absolutely had to. "In other words if we had to we could," he clarifies later. "I was trying to get the point across that our 1,400 acres could easily support our two families. My point being that we do not need to buy or rent more land."
4. This state statistic is somewhat misleading, as is the national average farm size of 434 acres per the *2012 Census of Agriculture*. Farms as small as a few acres count toward the total, which brings the average down. In truth, most midwestern grain farms are way over 1,000 acres. United States Department of Agriculture, *2012 Census of Agriculture*, volume 1, chapter 1: State Level Data, North Dakota, table 64, "Summary by Size of Farm," May 2014, https://www.agcensus.usda.gov/Publications/2012/Full_Report/Volume_1,_Chapter_1_State_Level/North_Dakota/st38_1_064_064.pdf.
5. Savory, *Holistic Management*, 424.
6. Economic gain for farmers is only the beginning. Bringing livestock back to the farm would revitalize sectors of rural economies that once thrived but are now dead. Family-run butcher shops, feed stores, livestock equipment stores, stores to retail the animal products, and jobs in helping farmers manage the animals are just a handful of examples of the effect livestock would have on rural communities.

7. Donald Schwert, "Fargo Geology: Why Is the Red River of the North So Vulnerable to Flooding?" North Dakota State University, first published in 2009, accessed December 15, 2014, http://www.ndsu.edu/fargo_geology/whyflood.htm.
8. "North Dakota Prairie: Our Natural Heritage," Northern Prairie Wildlife Research Center Online, May 5, 1999, http://www.npwrc.usgs.gov/resource/habitat/heritage/.

22. THE LIVESTOCK

1. I can't number the times angry cows have chased my father, usually over birthing- or calf-related intrusions on his part. Even when humans aren't around, cows go off by themselves to calve, away from the herd.

23. THE ALTERNATIVE TO HAY

1. Costello, *Prairie World*, 63.
2. Barber, *Third Plate*, 45.
3. "North Dakota Prairie: Our Natural Heritage," Northern Prairie Wildlife Research Center Online, May 5, 1999, http://www.npwrc.usgs.gov/resource/habitat/heritage/.
4. Nickens, "Vanishing Voices," 24.
5. Hay is best served the winter after its summer harvest, but it can last until the following winter if baled under the right conditions. Too-wet hay will mold, and too-dry hay will become unpalatable. Even properly harvested hay will break down—it is not filled with preservatives after all—and will lose some nutrients. Ranchers call hay from the previous year carry-over hay. They know it's not the highest-quality feed option, but it will do in a pinch.
6. Even Phil, with all his contiguous land, supplements some hay to the buffalo during the winter using the same bale-grazing method. It's probably impossible for Great Plains ranches to be entirely hay-free, but they can get very close, as Phil's and Gabe's operations are.
7. When he says "feeding hay" here, he means putting the cattle in pastures that contain preplaced bales, not starting the tractor and bringing them hay. He's talking about the number of days the cows are exposed to hay.

24. THE RESTORATION OF THE NATIVE PRAIRIE

1. Rogler and Lorenz, "Crested Wheatgrass," 91–93.

2. Roberts and Kallenbach, "Smooth Bromegrass," 1.
3. National Audubon Society, *Audubon's Birds and Climate Change Report*.
4. Nickens, "Vanishing Voices," 27.
5. Nickens, "Vanishing Voices," 27.
6. The Chinese ring-necked pheasant, an import from China, is South Dakota's state bird—oddly fitting in a state where non-native grasses like crested wheatgrass and smooth brome are pushing out native grasses.
7. Barber, *Third Plate*, 247.
8. Information about Brix readings in this paragraph and the next is from "Brix," Bionutrient Food Association, accessed November 12, 2014, http://bionutrient.org/bionutrient-rich-food/brix.
9. Barber, *Third Plate*, 61.
10. We can all test our food if we want to—refractometers are readily available online, easy to use, and not terribly expensive at around $100. My husband bought me one for Christmas, and I've been gleefully testing fruits and vegetables since.
11. Insects can actually smell sick plants with their antennae, which detect odor wavelengths emitted from plants. Barber, *Third Plate*, 54.
12. Not to be confused with the Egg Mobiles at Polyface Farm featured in Michael Pollan's *Omnivore's Dilemma*, which are more like tiny houses on wheels rather than modified livestock trailers.
13. For more on the human costs associated with megaslaughterhouses, see Eric Schlosser's *Fast Food Nation*.
14. Here I'm interviewing Gabe in August.

25. THE FARMERS' MARKET

1. Number calculated by taking total number of cows and calves in Perkins County (105,791 per the USDA's *2012 Census of Agriculture*, https://www.agcensus.usda.gov/Publications/2012/Online_Resources/County_Profiles/South_Dakota/cp46105.pdf) divided by the county's population (2,982 per the 2010 Census—https://www.census.gov/quickfacts/fact/table/perkinscountysouthdakota/HEA775216).
2. Barber, *Third Plate*, 15.
3. The phrase "high on the hog" refers to the costliest cuts of meat from a pig, taken from the back and upper leg. Wealthy people could afford to eat high on the hog, while poor people ate from the lower part, such as the belly, hocks, and feet. Barber, *Third Plate*, 152–53.

26. THE MESSAGE TO CONVENTIONAL FARMERS

1. Gabe is referring to agricultural tax incentives, some permanent and others temporary. During the 2014 tax year, for example, farmers could write off the cost of new machinery up to $250,000.
2. EQIP (the Environmental Quality Incentives Program) is sponsored by the Natural Resources Conservation Service.
3. Albritton, *Let Them Eat Junk*, 128–29.
4. Albritton, *Let Them Eat Junk*, 128–29.

BIBLIOGRAPHY

Albritton, Robert. *Let Them Eat Junk*. New York: Pluto Press, 2009.

Allred, B. W., S. D. Fuhlendorf, and R. G. Hamilton. "The Role of Herbivores in Great Plains Conservation: Comparative Ecology of Bison and Cattle." *Ecosphere* 2, no. 3 (2011): 1–17.

Ault, Toby R., Julia E. Cole, Jonathan T. Overpeck, Gregory T. Pederson, and David M. Meko. "Assessing the Risk of Persistent Drought Using Climate Model Simulations and Paleoclimate Data." *Journal of Climate* 27 (2014): 7529–49.

Barber, Dan. *The Third Plate: Field Notes on the Future of Food*. New York: Penguin Press, 2014.

Barnett, Barry J. "The U.S. Farm Financial Crisis of the 1980s." *Agricultural History* 74, no. 2 (2000): 366–80.

Benyus, Janine M. *Biomimicry: Innovation Inspired by Nature*. New York: HarperCollins, 2009.

Berry, Wendell. *The Unsettling of America: Culture and Agriculture*. San Francisco: Sierra Club Books, 1977.

Bowman, Maria S., and David Zilberman. "Economic Factors Affecting Diversified Farming Systems." *Ecology and Society* 18, no. 1 (2013): 33–47.

Buhler, Wayne. "Introduction to Insecticide Resistance." Pesticide Environmental Stewardship. October 23, 2014. http://pesticidestewardship.org/resistance/Insecticide/Pages/InsecticideResistance.aspx.

Burns, Catherine E., Scott L. Collins, and Melinda D. Smith. "Plant Community Response to Loss of Large Herbivores: Comparing Consequences in a South African and a North American Grassland." *Biodiversity and Conservation* 18 (2009): 2327–22.

Cain, Rita Marie. "Food, Inglorious Food: Food Safety, Food Libel, and Free Speech." *American Business Law Journal* 49, no. 2 (Summer 2012): 275–324.

Charles, Daniel. *Lord of the Harvest: Biotech, Big Money, and the Future of Food.* Cambridge MA: Perseus, 2001.

Conca, James. "It's Final—Corn Ethanol Is of No Use." *Forbes*, April 20, 2014. http://www.forbes.com/sites/jamesconca/2014/04/20/its-final-corn-ethanol-is-of-no-use/.

Costello, David F. *The Prairie World.* New York: Thomas Y. Crowell, 1969.

Dary, David. *Cowboy Culture: A Saga of Five Centuries.* New York: Knopf, 1981.

Davis, John H., and Ray A. Goldberg. *A Concept of Agribusiness.* Boston: Harvard University, 1957.

Dimitri, Carolyn, and Oberholtzer, Lydia. "Marketing U.S. Organic Foods: Recent Trends from Farms to Consumers." USDA Economic Research Service. Economic Information Bulletin no. 58 (September 2009).

D'Odorico, Paolo, Abinash Bhattachan, Kyle F. Davis, Sujith Ravi, and Christiane W. Runyan. "Global Desertification: Drivers and Feedbacks." *Advances in Water Resources* 51 (2013): 326–44.

Estabrook, Barry. *Tomatoland: How Modern Industrial Agriculture Ruined Our Most Alluring Fruit.* Kansas City: Andrews McMell, 2001.

Ewers, John C. *Plains Indian History and Culture: Continuity and Change.* Norman: University of Oklahoma Press, 1998.

Frankl, Viktor E. *Man's Search for Meaning.* Boston: Beacon Press, 2006.

Gassmann, Aaron J., Jennifer L. Petzold-Maxwell, Ryan S. Keweshan, and Mike W. Dunbar. "Field-Evolved Resistance to Bt Maize by Western Corn Rootworm." *PLOSONE* 6, no. 7 (2011): 1–7.

Gilbert, Elizabeth. *The Last American Man.* New York: Penguin Books, 2002.

Gladwell, Malcolm. *The Tipping Point: How Little Things Can Make a Big Difference.* New York: Back Bay Books/Little, Brown, 2000.

Hamilton, Lisa. *Deeply Rooted: Unconventional Farmers in the Age of Agribusiness.* Berkeley CA: Counterpoint, 2009.

Hart, John Fraser. *The Changing Scale of American Agriculture.* Charlottesville: University of Virginia Press, 2003.

Hoppe, Robert A. *Structure and Finances of U.S. Farms: Family Farm Report, 2014 Edition.* USDA Economic Research Service. Economic Information Bulletin no. 132, December 2014.

Hoppe, Robert A., and David E. Banker. *Structure and Finances of U.S. Farms: Family Farm Report, 2010 Edition*. USDA Economic Research Service. Economic Information Bulletin no. 66, July 2010.

Hribar, Carrie. *Understanding Concentrated Animal Feeding Operations and Their Impact on Communities*. Bowling Green OH: National Association of Local Boards of Health, 2010.

Huggins, David. R., and Reganold, John P. "No-Till: The Quiet Revolution." *Scientific American* 299, no. 1 (2008): 70–77.

Hurt, R. Douglas. *American Agriculture: A Brief History*. West Lafayette IN: Purdue University Press, 2002.

———. *The Problems of Plenty*. Chicago: Ivan R. Dee, 2002.

Ikerd, John. *Crisis and Opportunity: Sustainability in American Agriculture*. Lincoln: University of Nebraska Press, 2008.

Ingebritsen, S. E., Christopher McVoy, B. Glaz, and Winifred Park. "Florida Everglades: Subsidence Threatens Agriculture and Complicates Ecosystem Restoration." In *Land Subsidence in the United States: U.S. Geological Survey Circular 1182*, ed. Devin Galloway, David R. Jones, and S. E. Ingebritsen. Reston VA: U.S. Geological Survey, 1999.

Johnson, David, and Siobhan O'Connor. "These Charts Show Every Genetically Modified Food People Already Eat in the U.S." *Time*, April 30, 2015. http://time.com/3840073/gmo-food-charts/.

Karl, Thomas R., Jerry M. Melillo, and Thomas C. Peterson, eds. *Global Climate Change Impacts in the United States*. Cambridge UK: Cambridge University Press, 2009.

Kenner, Robert. "Exploring the Corporate Powers Behind the Way We Eat: The Making of *Food, Inc.*" In *Food, Inc.: How Industrial Food Is Making Us Sicker, Fatter, and Poorer—And What You Can Do About It*, ed. Karl Weber. New York: Public Affairs, 2009.

Koch, Adrienne, and William Peden. *The Life and Selected Writings of Thomas Jefferson*. New York: Random House, 1944.

Koohafkan, P., and B. A. Stewart, *Water and Cereals in Drylands*. London: Earthscan/The Food and Agriculture Organization of the United Nations, 2008.

Kramer, Mark. *Three Farms: Making Milk, Meat, and Money from the American Soil*. Boston: Little, Brown, 1980.

Leopold, Aldo. *A Sand County Almanac and Sketches Here and There*. New York: Oxford University Press, 1949.

Lim, J. Y., J. W. Yoon, and C. J. Hovde. "A Brief Overview of *Escherichia coli* O157:H7 and Its Plasmid O157." *Journal of Microbiology and Biotechnology* 20 no. 1 (2010), 5–14.

Mealer, Bryan. *Muck City*. New York: Crown Archetype, 2012.

National Audubon Society. *Audubon's Birds and Climate Change Report: A Primer for Practitioners*, Version 1.2. New York: National Audubon Society, September 2014.

National Research Council. *The Impact of Genetically Engineered Crops on Farm Sustainability in the United States*. Washington DC: National Academies Press, 2010.

Nickens, T. Edward. "Vanishing Voices." *National Wildlife* 48, no. 6 (October/November 2010): 22–29.

Obach, Brian K. "Theoretical Interpretations of the Growth in Organic Agriculture: Agricultural Modernization or an Organic Treadmill?" *Society and Natural Resources: An International Journal* 20, no. 3 (2007): 229–44.

Philpott, Tom. "A Brief History of Our Deadly Addiction to Nitrogen Fertilizer." *Mother Jones*, April 19, 2013. http://www.motherjones.com/tom-philpott/2013/04/history-nitrogen-fertilizer-ammonium-nitrate.

———. "Monsanto GM Soy Is Scarier Than You Think." *Mother Jones*, April 23, 2014. http://www.motherjones.com/tom-philpott/2014/04/superweeds-arent-only-trouble-gmo-soy.

Pollan, Michael. *The Omnivore's Dilemma: A Natural History of Four Meals*. New York: Penguin, 2006.

Pripps, Robert N. *The Big Book of Farmall Tractors*. Minneapolis: Voyageur Press, 2003.

Raloff, Jane. "Hormones: Here's the Beef: Environmental Concerns Reemerge over Steroids Given to Livestock." *Science News* 161, no. 1 (2002): 10–12.

Reed, Matthew. *Rebels for the Soil: The Rise of the Global Organic Food and Farming Movement*. Washington DC: Earthscan, 2010.

Roberts, Craig, and Robert L. Kallenbach. "Smooth Bromegrass." University of Missouri–Columbia Extension Publication, May 2000. http://extension.missouri.edu/p/g4672.

Roberts, Paul. *The End of Food*. New York: Houghton Mifflin Harcourt, 2008.

Robin, Marie-Monique. *The World according to Monsanto: Pollution, Corruption, and the Control of the World's Food Supply*. New York: New Press, 2010.

Rodriguez, Arturo. "Cheap Food: Workers Pay the Price." In *Food, Inc.: How Industrial Food Is Making Us Sicker, Fatter, and Poorer—And What You Can Do About It*, ed. Karl Weber. New York: Public Affairs, 2009.

Rogler, George A., and Russell L. Lorenz. "Crested Wheatgrass—Early History in the United States." *Journal of Range Management* 36, no. 1 (1983): 91–93.

Saunders, Steven, Charles Montgomery, Tom Easley, and Theo Spencer. *Hotter and Drier: The West's Changed Climate*. New York: Rocky Mountain Climate Organization and the Natural Resources Defense Council, March 2008.

Savory, Allan. *Holistic Management: A New Framework for Decision Making*. Washington: Island Press, 1999.

Schlebecker, John T. *Whereby We Thrive: A History of American Farming, 1607–1972*. Ames: Iowa State University Press, 1975.

Schlosser, Eric. *Fast Food Nation: The Dark Side of the American Meal*. Boston: Mariner Books/Houghton, Mifflin, Harcourt, 2012.

Sulc, R. Mark, and Benjamin F. Tracy. "Integrated Crop-Livestock Systems in the U.S. Corn Belt." *Agronomy Journal* 99 (2007): 335–45.

Tihansky, Ann B. "Sinkholes, West-Central Florida." In *Land Subsidence in the United States: U.S. Geological Survey Circular 1182*, ed. Devin Galloway, David R. Jones, and S.E. Ingebritsen, 95–106. Reston VA: U.S. Geological Survey, 1999.

Tober, Dwight A., and A. Dean Chamrad. "Warm-Season Grasses in the Northern Plains." *Rangelands* 14.4 (1992): 227–30.

Vose, R. S., D. R. Easterling, K. E. Kunkel, A. N. LeGrande, and M. F. Wehner. "Temperature Changes in the United States." In *Climate Science Special Report: Fourth National Climate Assessment*, vol. 1, eds. D. J. Wuebbles, D.W. Fahey, K.A. Hibbard, D. J. Dokken, B. C. Stewart, and T. K. Maycock. Washington DC: U.S. Global Change Research Program, 2017. pp. 185-206. https://science2017.globalchange.gov/chapter/6/.

Walker, Travis S., Harsh Pal Bais, Erich Grotewold, and Jorge M. Vivanco. "Root Exudation and Rhizosphere Biology." *Plant Physiology* 132 (2003): 44–51.

Welch, James, with Paul Stekler. *The Battle of Little Big Horn and the Fate of the Plains Indians*. New York: Penguin, 1995.

Wilcher, Robert. "The Greening of Milton Criticism." *Literature Compass* 7, no. 11 (2010): 1020–34.

Wilde, Matthew. "Study: Soil Eroding Faster than Estimated." *Cedar Valley Business Monthly Online*. May 22, 2011. http://wcfcourier.com/business/local/study-soil-eroding-faster-than-estimated/article_90a7f82c-8315-11e0-82f0-001cc4c03286.html.

Wolfe, David. *Tales from the Underground: A Natural History of Subterranean Life*. New York: Perseus, 2001.

Wozniak, Frank E. *Irrigation in the Rio Grande Valley, New Mexico: A Study and Annotated Bibliography of the Development of Irrigation Systems*. Fort Collins, CO: USDA, Forest Service, Rocky Mountain Research Station, 1998.

Wright, A. L., and E. A. Hanlon. "Organic Matter and Soil Structure in the Everglades Agricultural Area." University of Florida IFAS Extension, publication SL-301 (June 2009; revised January 2013).